Introduction to Phase Diagrams
in
Materials Science and Engineering

Introduction to Phase Diagrams
in
Materials Science and Engineering

Hiroyasu Saka

Aichi Institute of Technology, Japan

World Scientific

NEW JERSEY · LONDON · SINGAPORE · BEIJING · SHANGHAI · HONG KONG · TAIPEI · CHENNAI · TOKYO

Published in 2019 by

World Scientific Publishing Co. Pte. Ltd.
5 Toh Tuck Link, Singapore 596224
USA office: 27 Warren Street, Suite 401-402, Hackensack, NJ 07601
UK office: 57 Shelton Street, Covent Garden, London WC2H 9HE

Library of Congress Control Number: 2019948668

British Library Cataloguing-in-Publication Data
A catalogue record for this book is available from the British Library.

Originally published in Japanese as 坂公恭『材料系の状態図入門』
(Zairyoukei no Joutaizu Nyumon)
This English translation is published by arrangement with
Asakura Publishing Company Ltd., Tokyo, Japan.

INTRODUCTION TO PHASE DIAGRAMS IN MATERIALS SCIENCE
AND ENGINEERING

ISBN 978-981-120-370-1

For any available supplementary material, please visit
https://www.worldscientific.com/worldscibooks/10.1142/11368#t=suppl

Preface

Phase diagrams are a MUST for materials scientists and engineers (MSEs). However, understanding phase diagrams is a hard task for most MSEs. The audience of this book is two-fold. The first group is young MSEs, who start learning phase diagrams and are supposed to become specialists as MSEs. The other group are those who were trained in fields other than materials science and engineering but engage in research and/or development of materials after they are employed.

For the first group it is crucial to get accustomed to handling "diagrams (often very complex)" of "phase diagram", which they had never come across in their previous career. The second group must cope with phase diagrams by literally "on-the-job training". In this book the essence of phase diagrams is presented in a reader-friendly manner.

Phase diagram is based on thermodynamics, or rather thermodynamics itself. Therefore, phase diagram is to be taught based on thermodynamics in principle. However, for both beginners and on-the-job trainees this teaching method is too difficult a task. Elements of thermodynamics necessary to understand phase diagrams is briefly described in Chapter 3, however, this chapter can be skipped, since it is not necessarily crucial in order to master phase diagrams.

Ternary phase diagrams in Chapter 4 is far more complex than binary phase diagram. For this reason, a ternary phase diagram is nowadays less and less taught. However, in ceramics and semiconductors the ternary phase diagrams become more and more important. Recent software provides necessary information to handle ternary phase diagrams. However, needless to say, without fundamental knowledge of ternary phase diagrams it is impossible to *understand* ternary phase diagrams correctly. In this book ternary phase diagrams are presented in a completely original way, with many diagrams in full colour. This book is expected to be a *bible* for MSEs.

Hiroyasu Saka
Nagoya, Japan
August 2019

Contents

Chapter 1

Equilibrium Phase Diagrams

1.1 General Theory

1.1.1 Metallography and phase diagram

Metals are very useful. This is obvious if one looks back on the rise and development of human civilization. It is when human beings, after having experienced the stone age and bronze age, finally reached the iron age that modern civilization flourished. Figure 1.1 shows the production of steel all over the world. We are living in the middle of the iron age.

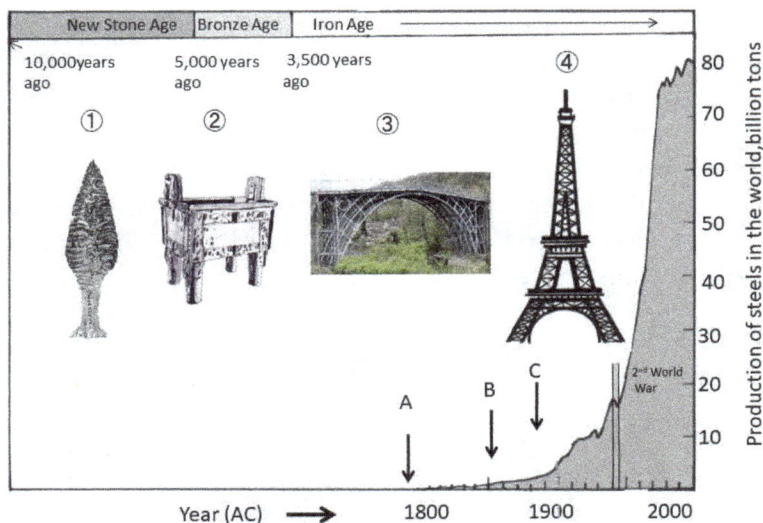

Fig. 1.1 History of production of iron (steel). ① Dagger in the new stone age. ② Bronze in old China (Yin). ③ Iron bridge (UK). ④ Eiffel tower (France). A: Iron bridge built (1781). B: Henry Bessemer invented a converter (1856). C: Eiffel tower built (1899).

Metals and alloys are backbones which support modern civilized societies because of their variety and flexibilities. Our ancestors learned diversity of metals and alloys by experience and made most of it, the compilation of which is metallurgy. Metallurgy is now indispensable to control microstructure of not only metallic materials but also ceramics and semiconductors.

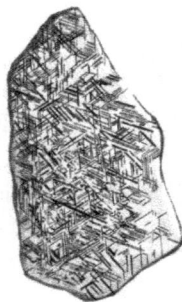

What is the microstructure of metals and alloys? Till the 18th century metals were not considered as crystals. In the early 19th century it was found that when a meteolite (usually composed of Fe and Ni) was lightly etched with a nitric acid, a pattern such as that shown in Fig. 1.2 appeared. Such a pattern was named Widmanstätten pattern after the discoverer Alois von Beckh Widmanstätten. This confirmed that metals and alloys are crystals just like minerals. Later this was confirmed by X-ray diffraction.

Fig. 1.2 Widmannstäten pattern.

Many people may think of quartz or diamond when they picture crystals to themselves (Fig. 1.3(a)). Internal atoms or molecules in these mineral crystals align themselves along a single orientation over the whole volume and are referred to as single crystals. Metallic crystals are different. Ordinary metallic materials are composed of aggregates of many smaller single crystallites (grains) with different orientations; this form is referred to as polycrystal (Fig. 1.3(b)). If the surface of a metal (etched chemically) is observed under an optical microscope, individual grains are revealed. A boundary between grains is referred to as a grain boundary. When all of the individual grains comprising a polycrystal have the single crystal structure, the material is referred to as a single-phase alloy. By contrast when a polycrystal is composed of grains with different crystal structures and/or

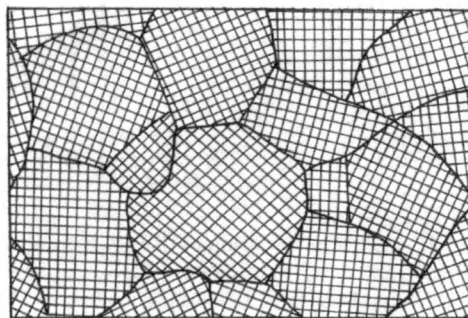

(a)   (b)

Fig. 1.3   (a) Single crystal (quartz). (b) Schematic illustration of a polycrystal.

composition, it is referred to as a multi-phase alloy. Morphology, size, distribution etc. of aggregates are referred to as microstructure. Properties of metals and alloys depend strongly on their microstructure, and the science to study factors controlling the relationship between microstructure and properties is referred to as metallography. Needless to say, this can be applied to materials in general.

The microstructure depends not only on the kind and the number of components (for instance Fe-Cr or Fe-Cr-Ni) comprising an alloy under consideration but also on its chemical composition (for instance, Fe–12%Cr versus Fe–18%Cr) and history such as temperature and heat-treatments which the alloy has undergone.

A map to indicate how the microstructure of a particular alloy under investigation depends on temperature is a phase diagram, where the equilibrium states are illustrated as functions of temperature, pressure and chemical composition.

1.1.2 Terminology

In metallography and phase diagrams many technical terms are used. Among them, the essential terms necessary to read this book are summarized in the following:

a. Alloy (alloy system)

A mixture that is created by adding other element(s) to a pure metal. The alloying elements can be metal, metalloids and non-metallic elements. Examples are Fe-C alloy (system), Au-Ag-Cu alloy (system) etc. Alloy and system are synonyms.

b. Component

How many kinds of elements are added to make an alloy?
Two-component system—binary system
Three-component system—ternary system
More than three-component system—multi-component system

c. Chemical composition, concentration

Chemical component indicates elements comprising an alloy. Relative amounts of individual elements are referred to as chemical composition or simply composition, usually indicated by percentage, either in weight percent (wt.%, w/o), mass percent (mass %) or atomic percent (at.%, a/o).

Weight concentration indicates the weight of each component (alloying elements) by percentage. Atomic concentration indicates the numbers of atoms by percentage. Needless to say, when one makes an alloy, a weight concentration is used. However, in theoretical calculation an atomic concentration is more convenient.

In the binary system, weight (mass) concentration is converted to atomic concentration using Eq. (1.1), where M_x, A_x and x stand for weight (mass)%, atomic % and the atomic weight of an element X, respectively. Those corresponding to the other element Y are M_y, A_y and y.

$$A_x = \frac{100M_x}{M_x + \frac{x}{y}(100 - M_x)} = \frac{100}{1 + \frac{x}{y}(\frac{100}{M_x} - 1)} \tag{1.1a}$$

$$M_x = \frac{100A_x}{A_x + \frac{y}{x}(100 - A_x)} = \frac{100}{1 + \frac{y}{x}(\frac{100}{A_x} - 1)} = \frac{100A_x x}{A_x x + 100y - A_x y} \tag{1.1b}$$

d. Phase

A phase has the same structure everywhere and is separated from any other phase by a surface, the *interface*. The three common states of matter are a gas, a liquid (melt) and a solid (crystal) (Fig. 1.4). A gas is always a single phase, however, this book treats only a solid and a liquid (condensed system), so that in the following a gas is not considered. A liquid can be a single phase (for instance, water + alcohol) or a dual phase (for instance, water + oil).

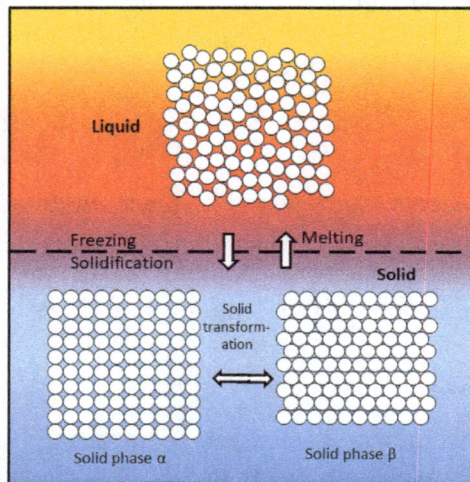

Fig. 1.4 Phase transformation.

In a solid (crystal), each phase has different crystal structure. Most important crystal structures are **BCC** (body centered cubic), **FCC** (face centered cubic) and **HCP** (hexagonal close packing). Particular phases are often denoted by Greek letters such as α-phase, β-phase, ..., Γ-phase. Figure 1.5 shows unit cells of BCC, FCC and HCP structures.

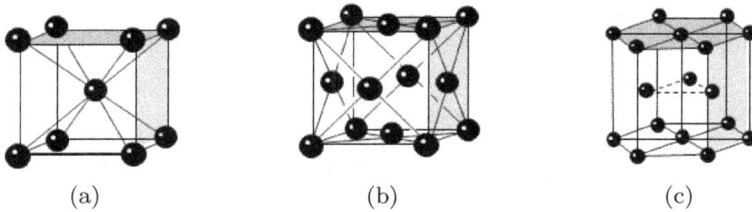

(a) (b) (c)

Fig. 1.5 Unit cells of BCC (a), FCC (b) and HCP (c).

The change from one phase to another is referred to as a phase transformation.

Example
liquid $\to$ solid (freezing or solidification)
solid $\to$ liquid (melting)

A transformation in the solid state in a pure element is referred to as an allotropic transformation.

Example
Fe

$$\text{BCC}(\alpha)\text{—}910°\text{C—FCC}(\gamma)\text{—}1390°\text{C—BCC}(\alpha)$$

This allotropic transformation is the base of heat treatments of iron and steel (quenching, tempering, etc.).

Sn

$$\text{Sn}(\alpha)\text{———}13°\text{C————Sn}(\beta)$$

This allotropic transformation accompanies a remarkable volume change. As a result, Sn is broken into pieces. This phenomenon is called the tin pest. The equilibrium temperature where the allotropic transformation takes place is 13°C, however, in practice it takes place at very low temperatures such as $-30°$C.

e. Magnetic transformation

Ferromagnetism at lower temperatures disappears at high temperatures, where the atomic arrangement remains unchanged and only the spin of electrons changes. Thus, strictly speaking, this transition is not a phase transformation, but this transition is so important that it is referred to as a magnetic transformation and denoted by a broken line in the phase diagram. The temperature above which magnetism disappears is referred to as the Curie temperature or the Curie point (Fig. 1.6).

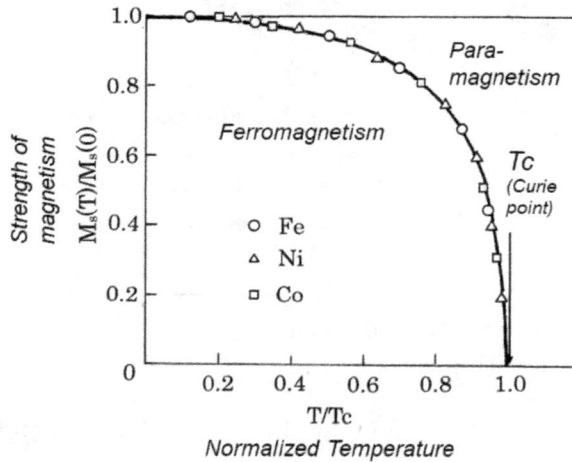

Fig. 1.6 Magnetic transformation.

f. Solid solution

Let us consider the case where an alloying element (i.e., solute) is added to a pure metal (i.e., solvent). While the amount of solute is small, they are dissolved into solvent completely. Suppose that solvent is coffee and solute is sugar. When a small amount of sugar (solute) is added to coffee (solvent) they are mixed up completely (Fig. 1.7(a)). This is referred to as a solution. Similar phenomenon takes place in solid state: This is referred to as a solid solution.

There are two kinds of solid solution: One is a substitutional type and the other is an interstitial type. In the former solvent atoms are replaced by solute atoms, while in the latter solute atoms enter one of the vacant spaces between solvent atoms. Thus, only elements with quite small atomic radii form an interstitial solid solution (Fig. 1.8).

Fig. 1.7 (a) Hot coffee. Sugar (solute) is completely dissolved in coffee (solvent). (b) Cooled coffee. Sugar is deposited: Solute (sugar) and solvent (coffee) are separated.

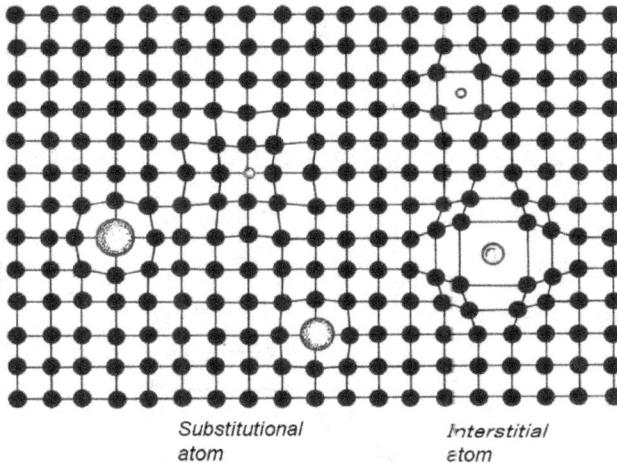

Substitutional   Interstitial
atom   atom

Fig. 1.8 Solid solution.

Example

H(0.64Å), B(0.97Å), C(0.77Å), N(0.71Å), O(0.60Å). Here, figures in parenthesis stand for the atomic radii, and 1Å(= 0.1 nm = 10^{-8} cm) corresponds to atomic radius, roughly speaking.

The maximum concentration to which a solute can enter as a solid solution is referred to as solubility or solubility limit. In the case of coffee, adding too much sugar results in deposition of sugar (Fig. 1.7(b)). The maximum amount of sugar that can be dissolved in coffee is its solubility limit. The same is true in the case of solid solution. In general, solubility limit increases with temperature. Needless to say, there are cases where the solubility limit does not change or even decreases with increasing temperature.

g. Ordered alloy, superlattice

An alloy where atoms of metal A and metal B arrange themselves in a regular manner is referred to as an ordered alloy. The crystal lattice of such an ordered alloy is referred to as a superlattice. Figure 1.9 shows some typical superlattices. (a) is based on BCC and referred to as B2 or $L2_0$. (b) is based on FCC and referred to as $L1_2$ or Cu_3Au-type. (c) is also based on FCC and referred to as $L1_0$ or CuAuI type. (d) is referred to as DO_3 or Fe_3Al type; it is based on BCC, composed of eight BCC-unit cells. B atom is surrounded by eight unlike atoms. In Heusler alloy, a ferromagnetic alloy, lattice sites ① and ③ are occupied by Cu, ② by Al, ④ by Mn. (e) shows a long period ordered alloy referred to as CuAuII type. This is based on $L1_0$ superlattice composed of ten $L1_0$ superlattices. Along the x axis, five $L1_0$ superlattices are connected, followed by another five $L1_0$ superlattice, but with Cu atoms exchanged with Au atoms.

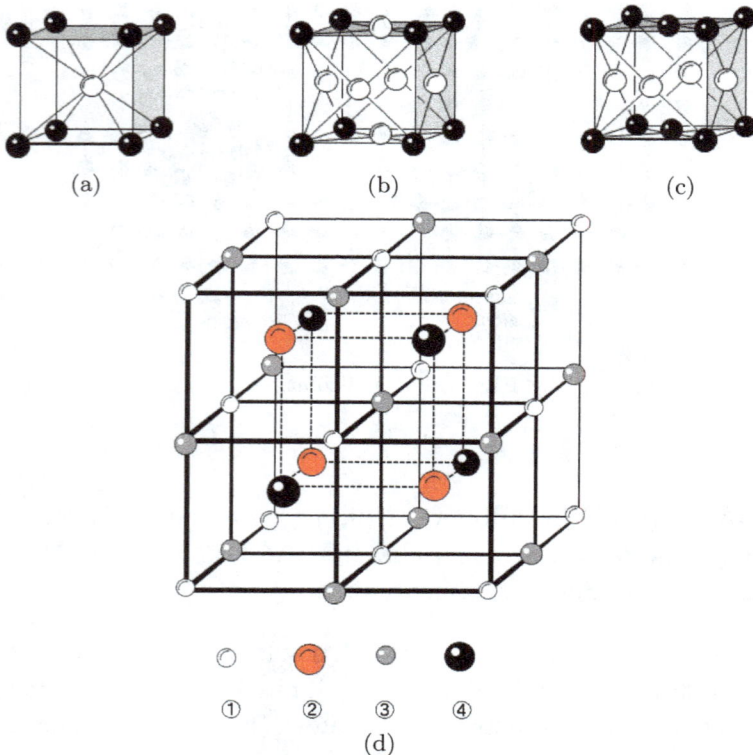

(a)   (b)   (c)

①   ②   ③   ④

(d)

Fig. 1.9 Superlattices. (a) B2, (b) $L1_2$, (c) $L1_0$, (d) DO_3 and (e) CuAuII type long-period order.

Structure of (d)	Ordered alloy	Lattice site			
		①	②	③	④
DO_3	Fe_3Al	Fe	Al	Fe	Fe
$L2_1$	Cu_2MnAl (Heusler alloy*)	Cu	Al	Cu	Mn
B32	NaTl	Na	Na	Tl	Tl

*Heusler is composed of Cu, Mn and Al, all of which are nonmagnetic elements, yet Heusler is the only ferromagnetic alloy.

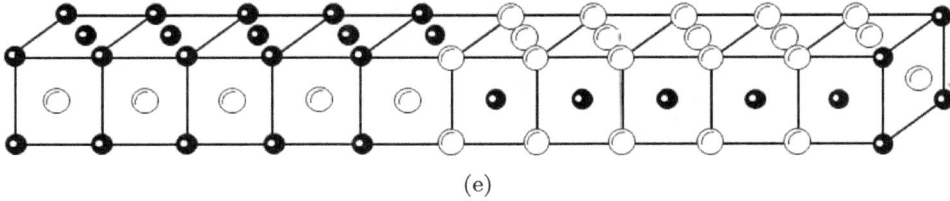

(e)

Fig. 1.9 (*continued*)

An ordered alloy transforms to a disordered one above a critical temperature. This is referred to as an order-disorder transformation (Fig. 1.10). The critical temperature above which an alloy becomes completely disordered is referred to as the Curie point in analogy to a magnetic transformation. When the Curie point is above the melting point of an alloy, the alloy is referred to as an intermetallic compound.

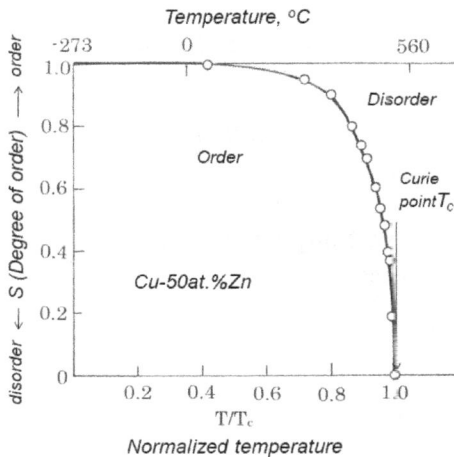

Fig. 1.10 Order-disorder transformation.

h. Equilibrium

The ultimate state where a system reaches eventually and stays permanently:
It can be compared to the terminal station of a railroad.

1.1.3 Phase rule of Gibbs

a. Equilibrium in a two-phase system

Supposing that two phases A and B are in contact, they are at equilibrium
with each other when the following three conditions are satisfied.
1) Mechanical equilibrium condition

$$P^A = P^B \text{ (pressure is equal)} \tag{1.2a}$$

2) Thermal equilibrium condition

$$T^A = T^B \text{ (temperature is equal)} \tag{1.2b}$$

3) Mass equilibrium condition

$$\mu_j^A = \mu_j^B \, (j = 1, 2, \ldots, c) \text{ (chemical potentials of each element } j \text{ are equal)} \tag{1.2c}$$

Here,

$$\mu_j^A = \left(\frac{\partial G^A}{\partial N_j} \right)_{T,V,N_i \neq j}$$

is the chemical potential of component j in phase A, where G^A is Gibbs free
energy of phase A, T is temperature and V is volume and similarly

$$\mu_j^B = \left(\frac{\partial G^B}{\partial N_j} \right)_{T,V,N_i \neq j}$$

is the chemical potential of component j in phase B.
    (See Sec. 3.1.1a for details.)

Chemical potential

The term "potential" may sound positive. However, it can be compared to the *unhappiness* (for the sake of simplicity let us make it *tax*) in human life. Phases (p) can be compared to nations and component (j) can be compared to people. Suppose that $j = 1$ is Japanese, $j = 2$ is American and $j = 3$ is Chinese etc. and that $p = 1$ is *Japan*, $p = 2$ is *United States of America* and $p = 3$ is *China*. If taxes imposed on Japanese ($j = 1$) living in Japan ($p = 1$), USA ($p = 2$) and China ($p = 3$) are equal ($\mu_1^1 = \mu_1^2 = \mu_1^3 = \cdots$), a net emigration of Japanese ($j = 1$) among these countries ($p = 1, 2, 3$) is zero. However, if tax imposed on Japanese $j = 1$ living in Japan $p = 1$ (chemical potential, μ_1^1) is heavier than that imposed on Japanese $j = 1$ living in USA $p = 2$ (chemical potential μ_1^2), in other words, $\mu_1^1 > \mu_1^2$, many Japanese would want to emigrate to the USA.

(Needless to say this is just a simile and motivation for emigration of people is much more complex.)

This painting, entitled "The Last of England", shows a dejected young couple setting out from home to start a new life overseas. It is clear from the sorrowful expression on their faces that they do not want to go; but economic and social condition in Europe during the 19th century forced emigration on millions of people just like them (cited from *Rainbow History Encyclopedia* published by Grisewood & Dempsey Ltd. in 1981). From the viewpoint of thermodynamics, the chemical potential of these people in Europe was higher than that in *New Worlds*.

b. Equilibrium in a multi-phase system and Gibbs' phase rule

Consider a system composed of p phases and c components at equilibrium. The degree of freedom f is given by

$$f = p(c-1) + 2 - c(p-1) = c - p + 2 \qquad (1.3)$$

Here, the degree of freedom is defined as the number of internal variables (pressure, temperature, chemical composition) which can be varied independently while satisfying the equilibrium conditions, Eqs. (1.2a), (1.2b) and (1.2c). Equation (1.3) is referred to as Gibbs[1] phase rule.

Proof. (This can be skipped without causing difficulty in understanding what follows.)

Suppose that the number of components is c and the number of phases is p.

1) Mechanical equilibrium condition

$$P^1 = P^2 = P^3 = \cdots = P^{(p)} = P \text{ (pressure of each phase is equal)} \quad (1.4)$$

2) Thermal equilibrium condition

$$T^1 = T^2 = T^3 = \cdots = T^{(p)} = T \text{ (temperature of each phase is equal)} \qquad (1.5)$$

3) Mass equilibrium condition

For example, the chemical potential of component *1* is equal in all of the phases $1 \sim p$.

$$\mu_1^1 = \mu_1^2 = \mu_1^3 = \cdots \mu_1^{(p)}$$

Thus, for component *1*, $(p-1)$ simultaneous equations are present. Similarly for components *2, 3, $\cdots$ c*,

$$\mu_2^1 = \mu_2^2 = \mu_2^3 \cdots = \mu_2^{(p)}$$

$$\mu_3^1 = \mu_3^2 = \mu_3^3 \cdots = \mu_3^{(p)}$$

$$- - - - - - - - - -$$

$$\mu_c^1 = \mu_c^2 = \mu_c^3 \cdots = \mu_c^{(p)} \qquad (1.6)$$

---

[1] Josiah Willard Gibbs (1839–1903). Theoretical physicist, professor of Yale University, USA. Gibbs published the phase rule in 1876, but its importance was not recognized till 1900.

must hold. The number of simultaneous equations is c $(p - 1)$. When an alloy composed of c components has p phases at equilibrium, the degree of freedom (f) is given by

$$f = p(c - 1) - c(p - 1) + 2 = c - p + 2.$$

Here, the third term 2 stands for pressure and temperature.

In general, in metals and alloys pressure can be ignored, so that

$$f = c - p + 1 \tag{1.7}$$

In what follows Eq. (1.7) will be used.

More concretely,

In one-component (unitary) system, $c = 1$. Therefore, $f = 2-$ p,

In two component (binary) system, $c = 2$. Therefore, $f = 3-$ p.

The significance of this is as follows:

Consider water as an example of a unitary system $(c = 1)$.[2] Under pressure of 1 atm two phases, i.e., water (liquid) and ice (solid) $(p = 2)$ can be coexistent only at the freezing point $(0°C)$. The temperature cannot be varied. Similarly, the temperature where water (liquid) and vapour (gas) can be coexistent, i.e., the boiling point, is 100°C and invariable. This situation has $f = 0$ and is referred to as an invariant system. By contrast, when only water (liquid) is present, $p = 1$ so that $f = 1$. In other words, water can exist in a range of temperature (between 0°C and 100°C).

For the invariant system $(f = 0)$ in a binary system $(c = 2)$, $p = 3$, that is, three phases coexist. The temperature and each of the chemical composition of the three phases are fixed. This state is an **invariant system.**

---

[2]The phase diagram of water is shown below.

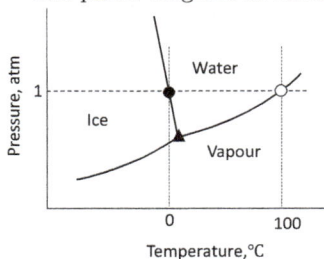

The triple point where ice, water and vapour coexist is shown by ▲ and at this point $f = 0$. Under 1 atm, ● is freezing point (0°C) and ○ is boiling point (100°C).

1.2   Binary Alloys in Equilibrium State

1.2.1   How to express

a. Single-phase region and two-phase region

For a binary alloy system, the kinds, relative amounts and chemical compositions of those phases that can coexist at an arbitrary temperature and composition are depicted by a phase diagram. Although the phase diagram itself says nothing about the distribution and morphology of each phase, in practice they can be inferred to some extent.

For a binary alloy $c = 2$ by definition, so that

$$f = 3 - p \tag{1.8}$$

The horizontal line shows the chemical composition denoted either by atomic percent, at.%, a/o or mass percent, mass% or equivalently weight percent, w%, w/o (see Sec. 1.1.2c).

Once the gross chemical composition of an alloy which one has prepared or is going to investigate is fixed, this particular alloy is represented by a vertical line passing through the gross chemical composition. Figure 1.11(a) shows the phase diagram of the binary Pb-Sn system.

Exercise.   In Fig. 1.11(a), (b), (c) the upper horizontal line denotes at.% and the lower one wt.% . Using Eq. (1.1a) calculate at.% of a 40wt.%Sn alloy and compare the result with Fig. 1.11(a).

Figure 1.11(a) shows a typical phase diagram which appears in common literatures. Information included in this diagram is as follows:

Pure metals
   Vertical lines at the two extremes correspond to pure metals (components). That is, the left vertical line indicates that Sn = 0%, i.e., pure Pb (Pb = 100%). Similarly, the right vertical line indicates that Sn = 100%, i.e. pure Sn.

   $T_{Pb} \cdots$ Melting temperature of Pb (327.5°C). This is a pure material, so that $c = 1$. At the melting temperature two phases (solid and liquid) coexist ($p = 2$). Therefore, according to Eq. (1.7), $f = 0$. In other words, the melting temperature is invariant and fixed.

   $T_{Sn} \cdots$ Melting temperature of Sn (232.0°C). For the same reason as T_{Pb}, $f = 0$. That is, the melting temperature is fixed (invariant).

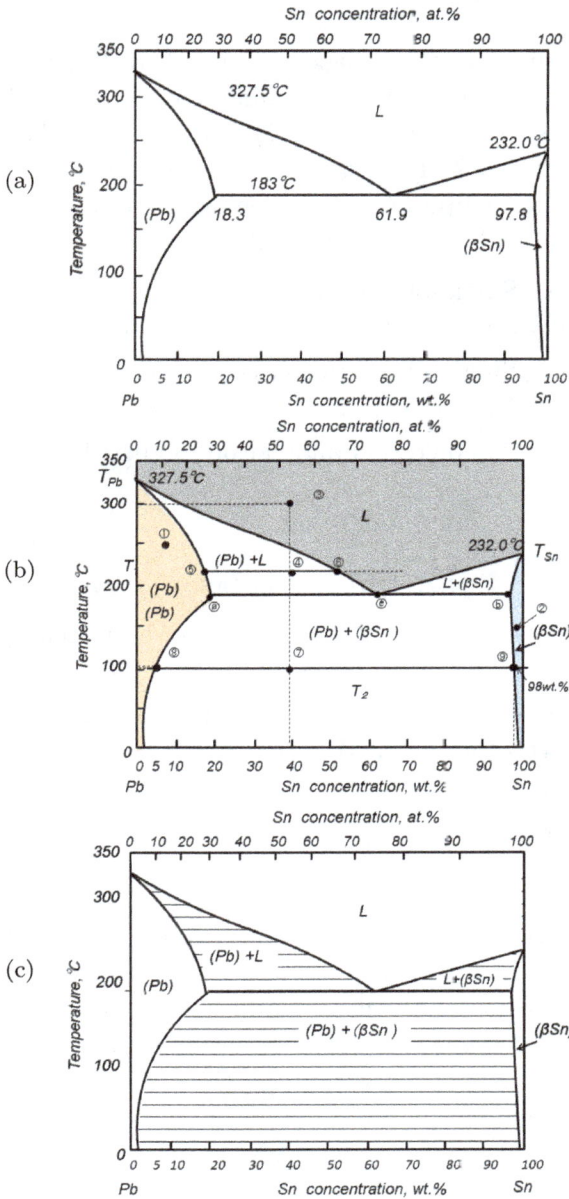

Fig. 1.11 (a) Phase diagram of the Pb-Sn binary system (reproduced with permission from Massalsiki[3]). (b) $T_{Pb} = 327.5°C$ is the melting point of Pb, $T_{Sn} = 232.0°C$ is the melting point of Sn. For the meaning of ① ~ ⑨, ⓐ, ⓑ and ⓒ, refer to the text. (c) Tie lines are added to (a).

---

[3]The most complete compendium of data on binary alloy systems; Massalski, T.B. and Okamoto, H. (eds.): *Binary Alloy Phase Diagrams*, ASM International, 1990.

Single-phase region

In the phase diagram those regions denoted by L, (Pb) and (βSn) indicate liquid phase, solid solutions mainly composed of Pb[4] and βSn (see Sec. 1.2.5), respectively. In Eq. (1.7) $c = 2$ (binary alloy), $p = 1$ (single phase), so that $f = 2$. That is, the number of independent internal variables is two, i.e. temperature and composition. In Fig. 1.11(b) regions of L, (Pb) and (βSn) are shaded and expanded two-dimensionally.

Point ① $\cdots$ Single-phase region consisting of Pb-rich solid solution (Pb) (10wt%Sn) at 250°C.

Point ② $\cdots$ Single-phase region consisting of Sn-rich solid solution (βSn) (99wt%Sn) at 150°C.

Point ③ $\cdots$ Pb-40wt.%Sn alloy at 300°C in a liquid phase (L).

Two-phase region

White regions in Fig. 1.11(b). For instance,

Point ④ $\cdots$ Mixture of (Pb) (20wt.%Sn, indicated by Point ⑤) and L phase (52wt.%Sn, indicated by Point ⑥) $f = 1$.

Point ⑦ $\cdots$ Mixture of (Pb) (5wt.%Sn, indicated by Point ⑧) and (βSn) (98wt.%Sn indicated by Point ⑨) $f = 1$.

Three-phase invariant system

Points along ⓐ–ⓑ (discussed later in detail).

Three phases, i.e., L (62wt.%Sn, indicated by Point ⓔ), (Pb) (20wt.%Sn, indicated by ⓐ) and (βSn) (96wt.%Sn, indicated by Point ⓑ) coexist. Since $p = 3$, Eq. (1.8) gives $f = 0$. That is, the degree of freedom is zero. In other words, neither temperature nor compositions of the coexisting three phases can be varied. This state is an invariant system.

b. Tie line and lever principle

In a two-phase region, the chemical compositions of two phases are represented by points where the horizontal line (representing the temperature under consideration) meet the boundaries of the respective single-phase regions. Such a horizontal line is referred to as a tie line. Customarily, tie lines

---

[4]In Massalski *et al.*, (Pb) stands for the Pb-terminated solid solution.

are not indicated in the phase diagrams (Fig. 1.11(a)). However, beginners
are recommended to draw tie lines in a two-phase region as indicated in
Fig. 1.11(c).

Example

The compositions of L and (Pb) phases coexisting at temperature T_1
are represented by ⑥ and ⑤, respectively. The compositions of (Pb) and
(βSn) phases coexisting at temperature T_2 are represented by ⑧ and ⑨,
respectively.

The amounts of coexisting two phases are given by the lever principle.

Example

In Fig. 1.12 suppose that an alloy whose gross composition is x lies in a
two-phase region of (Pb)+(βSn) at T_3 (in this particular case 100°C). Let
us denote the amount of α phase by $[\%\alpha]$ and that of β phase by $[\%\beta]$. The
ratio is given by $[\%\alpha]/[\%\beta] = (x_2 - x / x - x_1) = n/m$.

Proof. Let us show that an alloy composed of α phase with x_1 and β phase
x_2 with a ratio of $n:m$ has a gross composition of x.

x is an internal division of x_1 and x_2 with a ratio $m:n$.

$$(x - x_1):(x_2 - x) = m:n$$
$$n(x - x_1) = m(x_2 - x)$$
$$nx - nx_1 = mx_2 - mx$$
$$(m + n)x = mx_2 + nx_1$$
$$\therefore x = (mx_2 + nx_1) / (m + n)$$

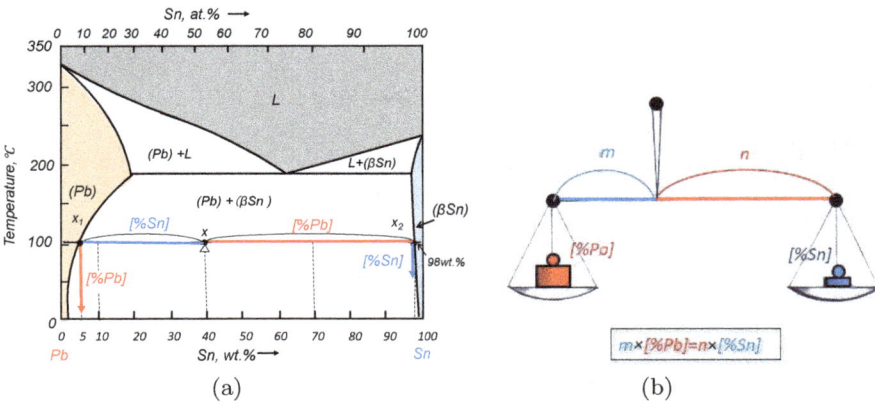

Fig. 1.12   Lever principle.

The reason why this relation is referred to as a lever principle is as follows. Let us consider a tie line $x_1 - x_2$ in Fig. 1.12 as a lever with a fulcrum at the gross composition of x of the alloy under consideration. Let us put a weight corresponding to $[\%\beta Sn]$ at one end of the lever x_2 and $[\%Pb]$ at the other end x_1. Then, in order for the lever to balance, the moment of the right half of the lever must be equal to that of the left. That is, $[\%Sn] \times (x - x_2) = [\%Pb] \times (x_1 - x)$. In other words, $[\%Sn] / [\%Pb] = (x - x_1 / x_2 - x)$.

Suppose that the gross chemical composition of an alloy x changes from the left (100wt.%Pb) to the right (100wt.%Sn) along x_1-x_2. When x is close to x_1 (10wt.%Sn in Fig. 1.12(a)), the alloy consists almost entirely of (Pb) phase with a composition of x_1. That is, $[\%Pb] \gg [\%Sn]$. In other words, $n \gg m$, as can be easily guessed. Inversely when x is very close to x_2, the alloy consists of entirely (βSn) phase, so that $[\%\beta Sn] \gg [\%Pb]$, i.e, $m \gg n$.

Exercise. Calculate [%Pb] and [%βSn] for alloys with x = 10wt.%Sn and 70wt.%Sn, respectively.

---

① Ends of a tie line (x_1 and x_2) indicate the chemical compositions of their adjacent single phases ((Pb) and (βSn)). Therefore, when x changes (moves) on a tie line x_1-x_2, the chemical compositions of two coexistent phases (Pb) and (βSn) remain unchanged.
② Amounts of (Pb) phase, i.e. [%Pb] and (βSn) i.e., [%βSn] change according to the lever principle.

These two rules are all that is necessary to understand the binary phase diagram. In what follows some simple examples will be presented.

1.2.2   Isomorphous (complete) solid solution

Two components (metals) are dissolved completely (Fig. 1.13(a)). Needless to say, the crystal structures of the two metals must be same and the lattice constants must be close to each other.

Example

Ag-Au, Cu-Ni, Ni-Pd, Cu-Au, etc.

One method of studying a phase diagram is to measure temperature change during cooling from a liquid as a function of time. This method is referred to as thermal analysis and the curve obtained as a cooling curve

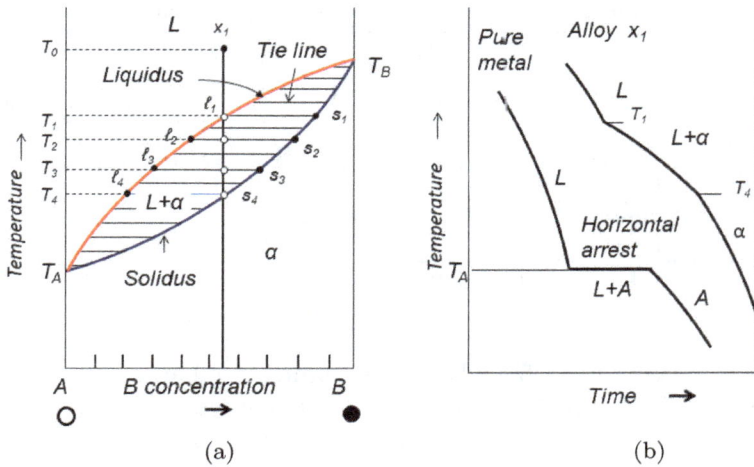

Fig. 1.13 (a) Solidification of an isomorphous solid solution. (b) Cooling curves.

(Fig. 1.13(b)). Needless to say, the temperature can be measured upon heating, and the curve obtained is referred to as a heating curve.

It is quite useful for understanding phase diagrams to consider cooling curves in *Gedanken Experiment*.

Let us start a cooling curve of a pure metal A (see Fig. 1.13(b)). Pure metal is a unitary system, so that the degree of freedom f is given by relation $f = 2 - p$.

Table 1.1 Solidification of pure metal A.

Temperature interval	Existing phase(s)	Degree of freedom: $f = 2 - p$
$T > T_A$	Liquid of pure metal A (L): single phase ($p = 1$)	$f = 1$ (temperature)
$T = T_A$	Solid of pure metal A freezes: two phases (liquid (L) and solid of pure metal A) ($p = 2$)	$f = 0$ (invariant system)
$T < T_A$	Solid of pure metal A: single phase ($p = 1$)	$f = 1$ (temperature)

At temperature T_A solidification starts, and two phases, i.e., L and A coexist. That is, $f = 0$, so that the temperature is fixed. This is referred to as a horizontal arrest. When solidification finishes, liquid (L) disappears, returning back to $f = 1$, and the temperature starts to decrease again.

Next, let us consider a cooling curve of an A-55at%B alloy.

Table 1.2 Solidification of an A-55at% B alloy.

Temperature interval	Existing phase(s)	Degree of freedom: $f = 3 - p$
$T > T_1$	Liquid (L): single phase ($p = 1$)	$f = 2$ (composition and temperature)
$T_1 > T > T_4$	L+solid solution (α): two phases ($p = 2$) "Freezing interval" α-phase crystallizes	$f = 1$ (temperature)
$T_2 > T$	α-phase: single phase ($p = 1$)	$f = 2$ (composition and temperature)

Solidification of an A-55at%B alloy is summarized as above. In the freezing interval ($T_1 - T_4$) the latent heat for solidification is evolved. As a result, the cooling rate is retarded, so that at T_1 and T_4 the cooling curves are bent.

Model experiment. Figure 1.14(a).

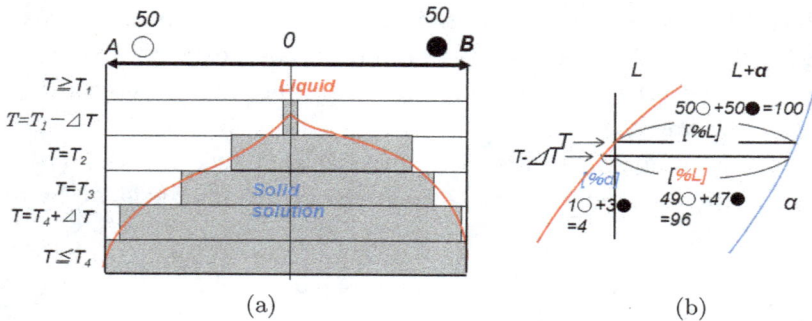

(a)   (b)

Fig. 1.14 Solidification of an isomorphous solid solution. (a) Breakdown of atoms. (b) Tie line at $T - \triangle T$ and lever principle.

Prepare 50 white balls and 50 black balls, where white balls represent metal A (denoted by ○) and black balls metal B (denoted by ●). Let us suppose that this ensemble consisting of 50 white balls and 50 black balls represents an A-50at.%B alloy.

In the liquid state all of the 50 white balls and 50 black balls are randomly mixed. Let us consider what happens on cooling this *alloy*.

Above T_1 is a single phase region of liquid phase (L). Therefore, all the 100 balls (both white and black) are in the liquid, the composition of which is 50at.%B. At temperature T_1 a solid solution (α) with composition s_1 (75at.%B) crystallizes from the liquid phase (L). The amount of α-solid solution is small. This can be visualized by drawing a tie line at a temperature slightly lower than T_1, that is, $T_1 - \triangle T (\triangle T \to 0)$ (Fig. 1.14(b)).

Table 1.3

Temperature	Liquid (L)			Solid solution (α)			Fig. 15
	Total number of atoms	Breakdown	Composition	Total number of atoms	Breakdown	Composition	
$T = T_1 + \Delta T$	100	50○+50●	50%B	0			(a)
$T = T_1 - \Delta T$	96	49○+47●	48.96%B	4	1○+3●	75%B	(b)
$T = T_2$	50	35○+15●	30%B	50	15○+35●	70%B	(c)
$T = T_3$	30	22○+8●	26.7%B	70	28○+42●	60%B	(d)
$T = T_4 + \Delta T$	4	3○+1●	25%B	96	47○+49●	51%B	(e)
$T = T_4 - \Delta T$	0			100	50○+50●	50%B	(f)

Distribution of A atoms (○) and B atoms (●) during solidification is schematically shown in Fig. 1.15.

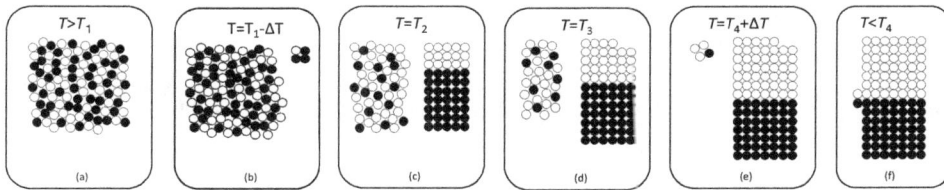

Fig. 1.15

Let us assume that the α-solid solution consists of one A atom (1○) and three B atoms (3●) with a chemical composition of A-75at.%B. The atoms remaining in (L) are now 49 A atoms and 47 B atoms (49○ + 47●), the chemical composition of which is A-48.96%B, slightly A-richer than the initial composition (A-50%B). In this way the composition of liquid follows a liquidus ℓ_1, ℓ_2, ℓ_3, ℓ_4 in Fig. 1.13(a) to A-richer side, and at the same time the composition of α-solid solution follows a solidus s_1, s_2, s_3, s_4, and eventually at T_4 all the atoms now belong to α-solid solution (solidification finished).

In general, when an alloy solidifies, a solid solution which crystallizes from the liquid phase has chemical composition different from that of the liquid phase.

1.2.3 Eutectic alloy

a. Component metals A and B have only negligibly small mutual solid solubility

Figure 1.16 shows an example. However small the mutual solubilities of A and B may appear, perfectly pure materials do not exist. The purest

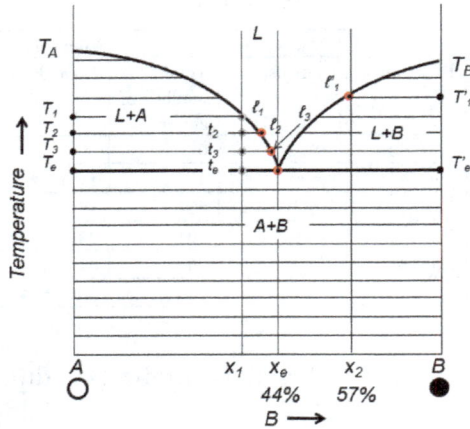

Fig. 1.16 Solidification of a eutectic alloy. T_A and T_B are the melting points of A and B, respectively. $T_e - T'_e$ is the eutectic temperature.

material available currently is Si, the purity of which is 99.999999999%. This is called 11N (eleven nine). This demonstrates that even the purest Si contains impurities, albeit very small.

Therefore, when the solubility of α-solid solution is negligibly small, the vertical line (indicating temperature), solidus and solid solubility curve (referred to as **solvus**) of the solid solution are so close that they overlap, giving an impression that they are one line, as shown in Fig. 1.17(a). However, they are independent lines when enlarged as shown in Fig. 1.17(b).

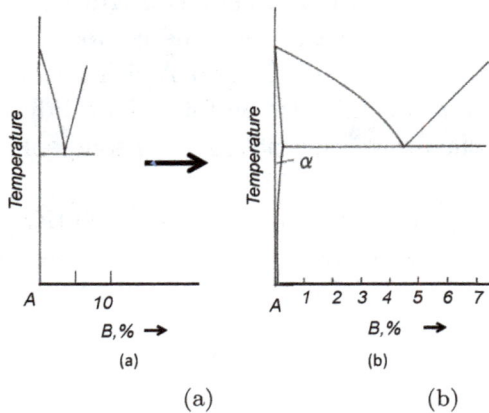

Fig. 1.17 Phase diagrams with negligibly small solubility limit. (a) Solubility of α phase is very small. (b) Temperature axis, solidus and solvus are overlapped.

Let us start from a eutectic composition ($x_e = 57\%B$) and consider what happens on solidification.

Table 1.4

⟨ Eutectic composition; $x_e = 57\%B$ ⟩		
Temperature interval	Existing phase(s)	Degree of freedom (f)
$T > T_e$ (eutectic temperature)	Liquid L (57%B): single phase ($p = 1$)	2 (composition and temperature)
$T = T_e$ (eutectic temperature)	$L+A+B$ (**eutectic reaction** $L \rightarrow A+B$ in progress) ($p = 3$)	0
$T < T_e$ (eutectic temperature)	A+B: 2 phases of solid ($p = 2$)	1 (temperature)

At $T = T_e$ three phases, i.e., L, A and B coexist, so that $f = 0$. This is a eutectic reaction which is an invariant reaction. Microstructure formed by a eutectic reaction consists of A and B phases intricated with each other as shown in Fig. 1.18(a). This type of microstructure is referred to as a eutectic structure. Figure 1.18(b) shows a transmission electron micrograph (TEM) of eutectic structure of an Al-Cu alloy.

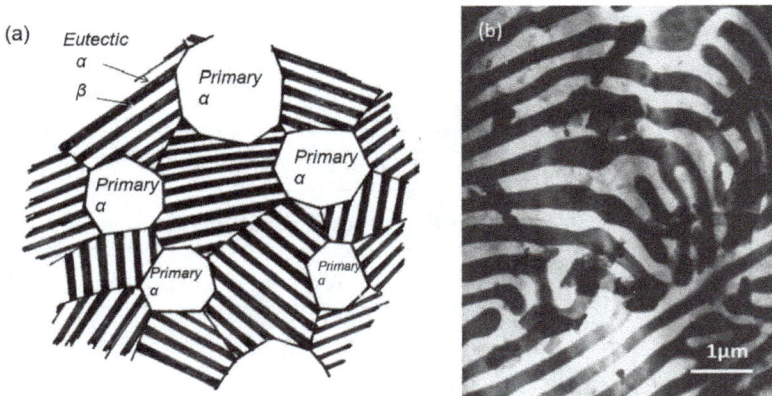

Fig. 1.18 (a) Schematic illustration of the primary (α) and eutectic structure. (b) TEM of eutectic structure in an Al-36wt.% Cu alloy. White areas are Al (FCC) and black areas are θ (CuAl$_2$) phase. For the phase diagram of Al-Cu binary system refer to Fig. 2.2.

Next, consider the case of an alloy with $x_1 = 44\%B$ (Fig. 1.19) (Table 1.5).

Table 1.5

⟨ Composition $x_1 = 44\%B$ ⟩		
Temperature interval	Existing phase(s)	Degree of freedom (f)
$T > T_1$	Liquid L (44%B): single phase ($p = 1$)	2 (composition and temperature)
$T_1 > T > T_e$ (eutectic temperature)	$L+A$: 2 phases (freezing interval) ($p = 2$)	1 (temperature)
$T = T_e$ (eutectic temperature)	$L+A+B$ (**eutectic reaction** $L \to A+B$ in progress) ($p = 3$)	0
$T < T_e$ (eutectic temperature)	$A+B$: 2 phases ($p = 2$)	1 (temperature)

Now let us consider this case in more detail in Table 1.6.

Table 1.6

	Liquid (L)		Solid (S)			
Temperature interval	Composition	Amount	Composition	Amount	Composition	Amount
T_1	$\ell_1(x_1)$	1		0		
T_2	ℓ_2	T_2t_2	pure A	$t_2\ell_2$		
$T_3 = T_e + \Delta T$	ℓ_3	T_3t_e	pure A	$t_e\ell_3$		
T_e	$\ell_e \to A+B$ (**eutectic reaction**) in progress					
$T_4 < T_e$		0	pure A	x_1B	pure B	x_1A

Model experiment. Figure 1.19.

Let the alloy consist of 50 A elements (○) and 40 B elements (●) (a total of 90 atoms), the composition of which is A-44%B.

To summarize, the distribution of A atoms (○) and B atoms (●) during solidification is schematically shown in Fig. 1.19.

Table 1.7

Temperature interval	Liquid (L)			Solid			Fig. 20
	Total amount of atoms	Breakdown	Composition	Total amount of atoms	Breakdown	Composition	
$T = T_1 - \Delta T$	90	50○+40●	44%B	0			(a)
$T = T_1 - \Delta T$	88	48○+40●	48%B	2	2○	100%A	(b)
$T = T_2$	80	40○+40●	50%B	10	10○	100%A	(c)
$T_3 = T_e + \Delta T$	70	30○+40●	57%B	20	20○	100%A	(d)
	Liquid with a eutectic composition			100%A (20○) (Primary solid)			
$T = T_e$	Eutectic reaction $(L \rightarrow A+B)$			↓			
$T = T_4 = T_e - \Delta T$	57%B (30○+40●) (Eutectic structure)			100%A (20○) (Primary solid)			(e)
	90 (50○+40●)						

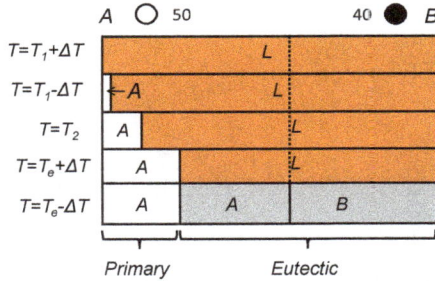

Fig. 1.19   Solidification of a eutectic alloy.

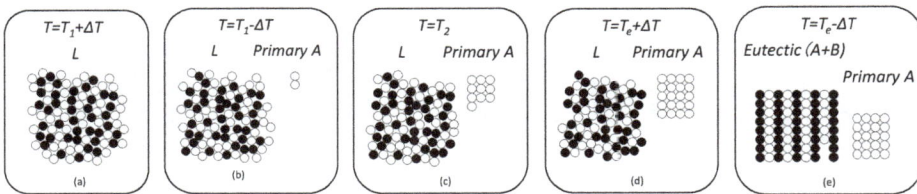

Fig. 1.20

A solid solution which crystallizes from a liquid is referred to as a primary solid solution. The primary grains are coarse while a eutectic structure is much finer (Fig. 1.18(a)).

Exercise. Perform similar analysis on an alloy x_2 in Fig. 1.16.

Hypoeutectic and hypereutectic alloys

Those alloys whose gross chemical compositions deviate from the eutectic composition are referred to as hypo- and hyper eutectic alloys.

Hypoeutectic $x < x_e$
Euectic $x = x_e$
Hypereutectic $x > x_e$

Exercise. Consider how the amount of eutectic structure depends on the concentration of B in Fig. 1.16.

Solution. Let the amount of eutectic structure be denoted by [% eutectic]. [% eutectic] is nothing but the amount of liquid remaining just above T_e (denoted by [%L_e]), so that [% eutectic] = [%L_e].

In Fig. 1.21,
for alloy x_1

$$[\% \text{ eutectic}] = [\%L_e] = T_e t_e / T_e \ell_e$$

for alloy x_e

$$[\% \text{ eutectic}] = [\%L_e] = 100\%$$

for alloy x_2

$$[\% \text{ eutectic}] = [\%L_e] = T'_e t'_e / T'_e \ell_e.$$

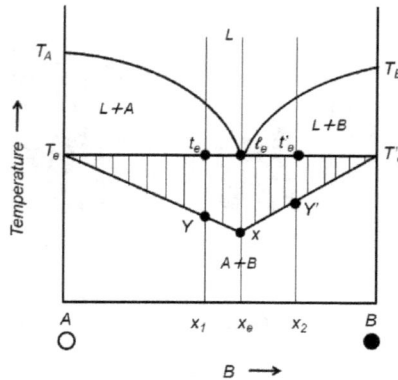

Fig. 1.21   Amount of eutectic structure.

Thus, the amount of eutectic structure ([% eutectic]) is 100% at the eutectic composition, decreasing proportionally on deviating from the eutectic composition.

A triangle connecting $T_e - X - T'_e$ in Fig. 1.21 shows the amount of eutectic structure, [% eutectic].

For alloy x_e

$$[\% \text{ eutectic}] = 100\% = \ell_e X.$$

For alloy x_1

$$[\% \text{ eutectic}] = [\% L_e] = T_e t_e / T_e \ell_e = t_e Y / \ell_e X$$

For alloy x_2

$$[\% \text{ eutectic}] = [\% L_e] = T'_e\, t'_e\, /\, T'_e\, \ell_e = t'_e\, Y'\, /\ell_e X$$

The vertical distance from the horizontal line $T_e - T'_e$ to a bent line $T_e - X - T'_e$ corresponds to [% eutectic].

To ensure this, let us calculate the total amount of A for an alloy x_1.

$$[\% A] = [\% \text{primary A}] + [\% \text{eutectic A}],$$

where [% eutectic A] stands for the amount of A in the eutectic structure.

$$
\begin{aligned}
[\% A] &= [\% \text{primary A}] + [\% \text{ eutectic A}] \\
&= t_e \ell_e / T_e \ell_e + [\% L_e] \times (\ell_e T'_e / T_e T'_e) \\
&= t_e \ell_e / T_e \ell_e + (T_e t_e / T_e \ell_e) \times (\ell_e T'_e / T_e T'_e) \\
&= \frac{T_e \ell_e - T_e t_e}{T_e \ell_e} + \frac{T_e t_e}{T_e \ell_e} \times \frac{T'_e \ell_e}{T_e T'_e} \\
&= 1 - \frac{T_e t_e}{T_e \ell_e} + \frac{T_e t_e}{T_e \ell_e} \times \frac{T'_e \ell_e}{T_e T'_e} \\
&= 1 - \frac{T_e t_e}{T_e \ell_e}\left(1 - \frac{T'_e \ell_e}{T_e T'_e}\right) \\
&= 1 - \frac{T_e t_e}{T_e \ell_e} \times \frac{T_e T'_e - T'_e \ell_e}{T_e T'_e} \\
&= 1 - \frac{T_e t_e}{T_e T'_e} \\
&= \frac{T_e T'_e - T_e t_e}{T_e T'_e} \\
&= \frac{t_e T'_e}{T_e T'_e}
\end{aligned}
$$

When a liquid alloy is cooled in a crucible, it is cooled from the wall, so that freezing starts from the wall towards the inside. Such a structure is referred to as columnar structure (Fig. 1.22(a)). When a liquid is solidified intentionally from one end, the eutectic structure aligns itself like a bamboo (Fig. 1.22(b)). Such a process is referred to as directional solidification.

Consider a liquid (melt), from which a single solid phase is formed on freezing. When it is cooled from one end, a single crystal can be grown. This method is referred to as Bridgman method. A tipped crucible is inserted in

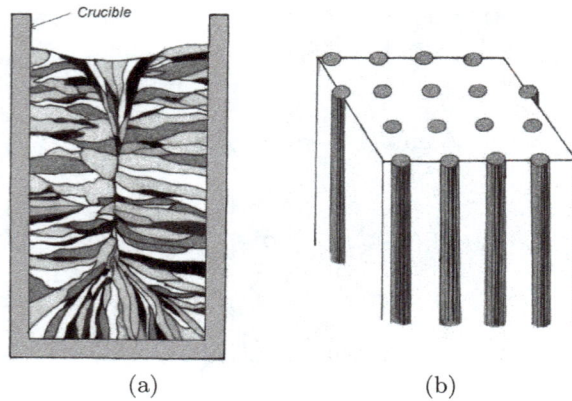

Fig. 1.22   (a) Columnar structure. (b) Directional solidification.

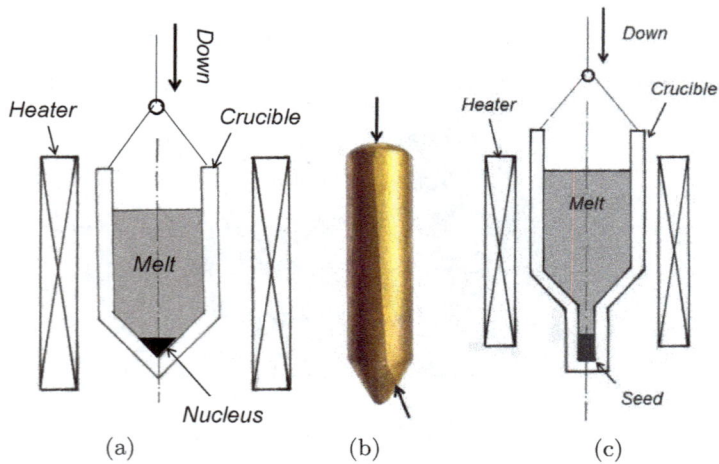

Fig. 1.23   (a) Bridgman method.   (b) Bi-crystal of Cu grown by Bridgman method (failure). Arrows show the grain boundary of the two crystals. (c) Bridgman method with a seed crystal.

a vertical furnace and slowly lowered. Then, solidification starts from the tip of the crucible, resulting in the growth of a single crystal (Fig. 1.23(a)). Figure 1.23(b) shows a Cu crystal grown by the Bridgman method. This crystal was designed to be a single crystal, however, two crystals are nucleated at the tip simultaneously, which were inherited upwards, resulting in the growth of a bi-crystal[5] consisting of two differently orientated grains. This is a failure. In order to avoid concurrent nucleation of more than one

---

[5]Note that a bicrystal is different from a twin (see Fig. 2.26).

(a)

(b)

(c)

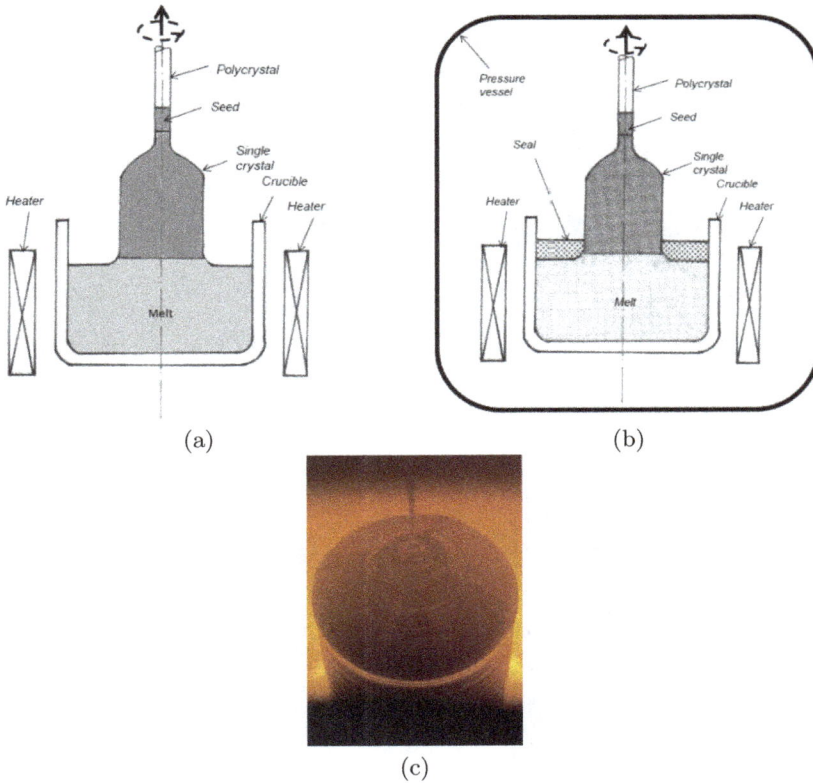

Fig. 1.24 (a) Czochralski method. (b) LEC method. (c) Photo of a growing Si in CZ method (courtesy of SUMCO Corporation).

grain, a seed crystal is inserted at the bottom of a crucible, the lower part of which remains a solid. Then, the crucible is lowered. The probability of growing a single crystal is much improved. On top of this, by selecting the orientation of the seed crystal, a single crystal with an intended orientation can be grown easily (Fig. 1.23(c)).

One of the most popular and important single crystals which we use in everyday life is Si. Si single crystals are grown by Czochralski method or floating-zone method. Si grown by the former method is called CZ-Si, and the latter FZ-Si. Significance of abbreviation of CZ and FZ is different. Outline of Czochralski (CZ) method is shown in Fig. 1.24(a)(c).

In the CZ method, a raw material is molten in a large crucible, into which a thin single crystal, i.e. a seed crystal is inserted. The molten raw material sticks to the seed which is pulled up slowly with rotation. When the raw material is volatile, the whole set-up is encapsulated in a high-pressure

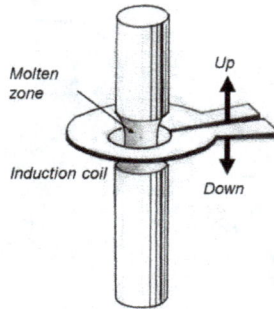

Fig. 1.25 Floating zone melting method.

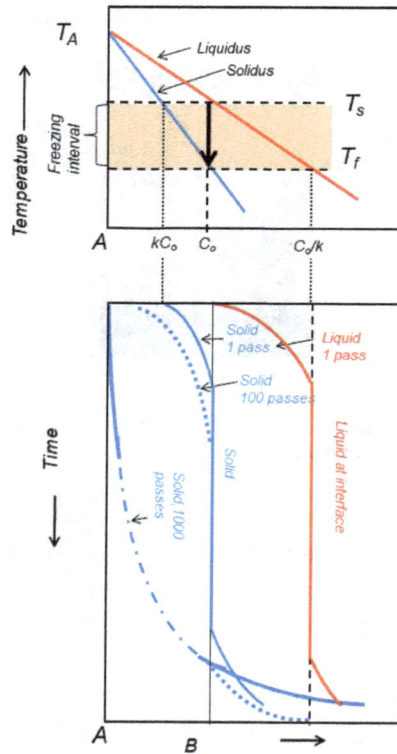

Fig. 1.26 Purification by zone melting.

vessel to prevent evaporation. This method is called a liquid encapsulated Czochralski (LEC) (Fig. 1.24(b)).

In the FZ method, a polycrystal rod stands vertically with a heater which heats locally. Usually an induction coil is used for radio-frequency heating

(Fig. 1.25).[6] When a part of the sample gets molten, the coil is pulled up (or down) slowly, resulting in solidification of the molten part into a single crystal. When the original polycrystal is swept completely with this operation, the original polycrystal becomes a single crystal. Since the molten zone is floating in between upper and lower solids, this method is referred to as **floating zone (FZ)** method. In this method there is no contact between a material to be treated and the crucible. Contamination from the crucible is not present, so that a high-purity material is obtained. When there is no contamination from a crucible, a lateral crucible (boat) is used and a heater is tranversed horizontally. This method is called a zone melting method. The floating zone method is a kind of zone melting method.

When a material containing a small amount of impurity is treated with FZ or ZM method, the impurity is swept to one end of the original sample, making the swept part purer (Fig. 1.26).

Suppose that the concentration of impurity in the liquid at T_s is C_o, then the corresponding impurity concentration in the solid is kC_o, which is lower than C_o. When temperature reaches T_f, the lower limit of the freezing zone $(T_s - T_f)$, the impurity concentration in the solid becomes C_o. Thus, the impurity concentration at the head of the sample gets lower. In other words, impurity is swept to the tail of the sample. By repeating this process, impurity level is reduced except for the tail. If the tail is cut off, the remaining part is purified.

b. Solubility is present in solids

This corresponds to the case in Fig. 1.17(b) where the solubility limits of α-phase and β-phase are significant (see Fig. 1.27).

Exercise. Draw schematically the cooling curves of the corresponding alloys shown in Fig. 1.27(a).

1.2.4 Eutectoid reaction

At high temperature liquid L solidifies into an isomorphous solid solution γ, which decomposes into $\alpha + \beta$ on further cooling. This reaction is just similar to a eutectic reaction except that all the phases involved (γ, α, β) are solids (Fig. 1.28).

---

[6]In addition to radio-frequency heating, electron-beam, infrared-concentrated oven are also used.

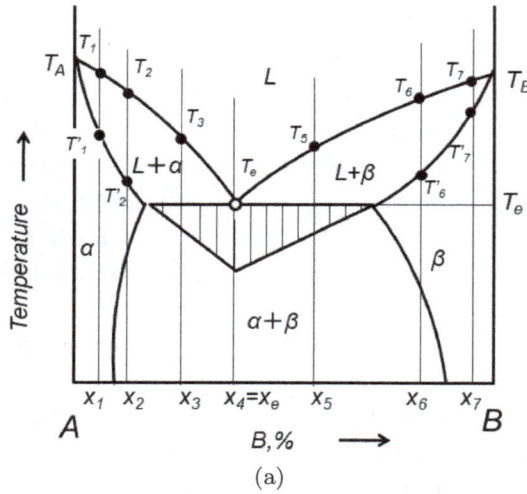

(a)

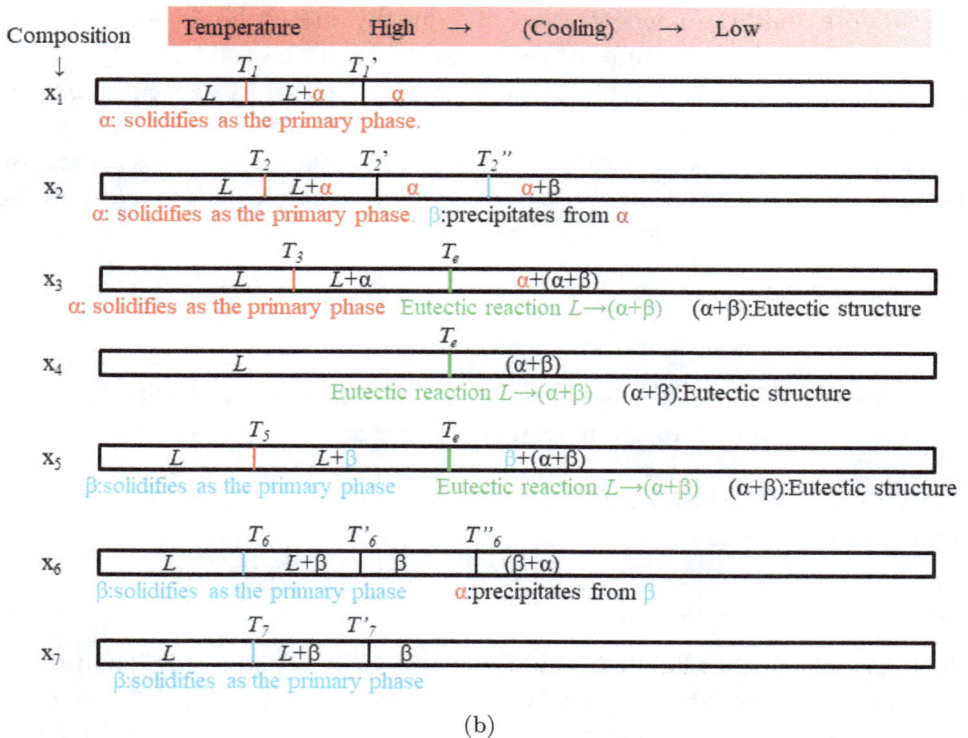

(b)

Fig. 1.27 Eutectic reaction in the binary system where the primary solid solutions have solubility. (a) Triangle below T_e where a eutectic reaction takes place indicates the amount of eutectic structure. (b) Reactions which occur when alloys (with contents) of x_1, x_2, x_3, x_4, x_5, x_6 and x_7 are cooled from liquid L.

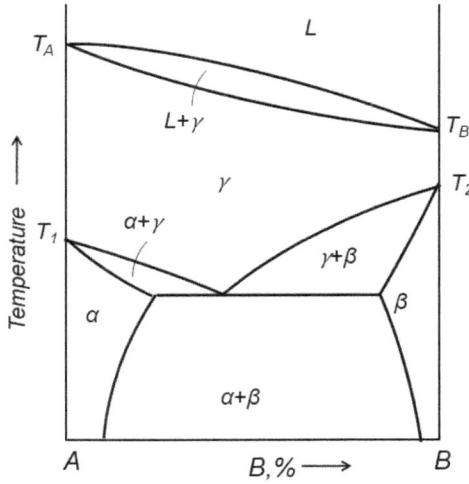

Fig. 1.28   Eutectoid reaction.

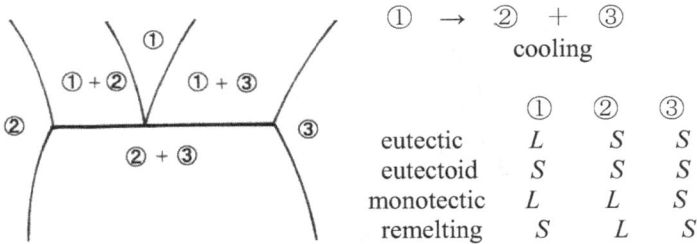

Fig. 1.29   Eutectic type reactions.

1.2.5   Other eutectic-type invariant reactions

Typical eutectic-type invariant reactions are summarized in Fig. 1.29.

1.2.6   Peritectic reaction

This is a reverse reaction to eutectic reaction. What happens on cooling corresponds exactly to what happens on heating a eutectic alloy (Fig. 1.30(a)(b)).

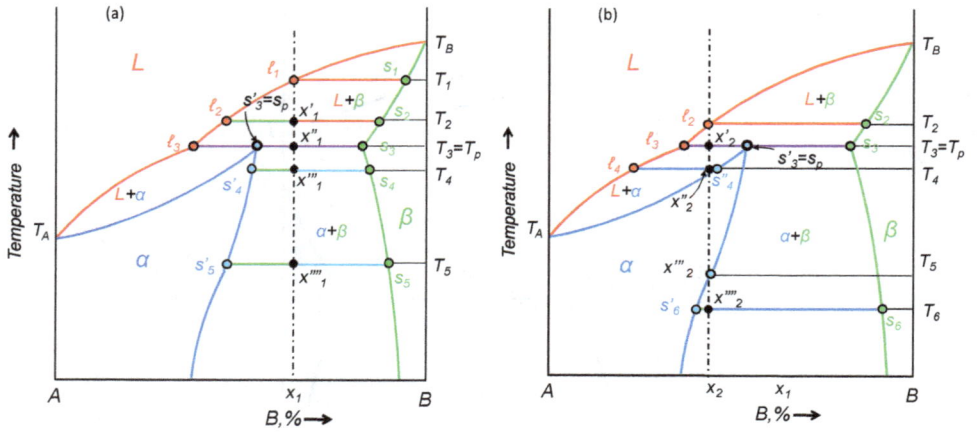

Fig. 1.30 (a)(b) Peritectic reaction.

The freezing process of an alloy x_1 is summarized in Table 1.8.

Table 1.8

Temperature	L Composition	L Amount	β Composition	β Amount	α Composition	α Amount
$T > T_1$	x_1	100%				
$T = T_1 - \Delta T$	$\ell_1 (\fallingdotseq x_1)$	$\sim 99\%$	s_1	$\sim 1\%$		
$T = T_2$	ℓ_2	$s_2x'_1$	s_2	$\ell_2 x'_1$		
$T = T_3 + \Delta T$	$\fallingdotseq \ell_3$	$\sim s_3x''_1$	$\fallingdotseq s_3$	$\sim \ell_3x''_1$		
$T = T_3$	$L(\ell_3) + \beta(s_3) \quad \rightarrow \quad \alpha(s'_3 = s_p)$ **Peritectic reaction**					
$T = T_3 - \Delta T$			s_3	$s'_3x''_1$	s'_3	x''_1s_3
$T = T_4$			s_4	$s'_4x'''_1$	s'_4	x'''_1s_4
$T = T_5$			s_5	$s'_5x''''_1$	s'_5	x''''_1s_5

The freezing process of an alloy x_2 is summarized in Table 1.9.

Table 1.9

Temperature	L Composition	L Amount	β Composition	β Amount	α Composition	α Amount
$T > T_2$	x_2	100%				
$T = T_2 - \Delta T$	$\ell_2(C)$	$\sim 99\%$	s_2	$\sim 1\%$		
$T = T_3 + \Delta T$	ℓ_3	$s_3x'_2$	s_3	$\ell_3x'_2$		
$T = T_3$	$L(\ell_3) + \beta(s_3) \quad \rightarrow \quad \alpha(s'_3 = s_p)$ **Peritectic reaction**					
$T = T_3 - \Delta T$	ℓ_3				$\sim s'_3$	$\sim \ell_3x'_2$
$T = T_4 + \Delta T$	ℓ_4	$\sim 1\%$			$s''_4 (\fallingdotseq x_2)$	$\sim 99\%$
$T = T_4 \sim T_5$					x_2	100%
$T = T_5 - \Delta T$			s_5	$\sim 1\%$	s'_5	$\sim 99\%$
			β precipitates from α.			
$T = T_6$			s_6	s'_6x_1	s'_6	x_2s_6

In order to understand the freezing process of a peritectic reaction let us perform a model experiment on an alloy x_1(50at.%B) in Fig. 1.31.

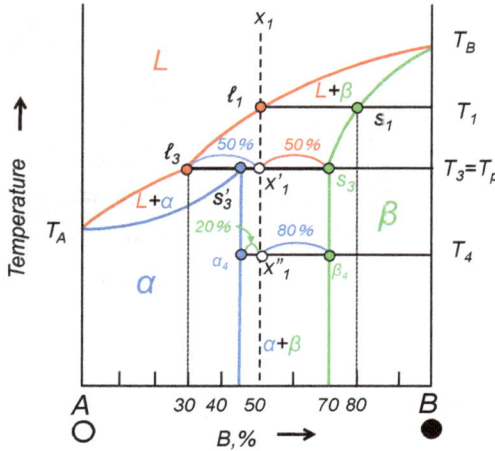

Fig. 1.31

Model experiment. Let us consider an alloy consisting of 50 A atoms (○) and 50 B atoms (●).

Table 1.10 $<x_1(=50at.\%B:50○ - 50●)>$

	L			α			β			Fig. 33
Temperature interval	Total number of atoms	Breakdown	Composition	Total number of atoms	Breakdown	Compositicn	Total number of atoms	Breakdown	Composition	
$T > T_1$	100	50○+50●	50%B							(a)
$T = T_1 - \varDelta T$	95	49○+46●	48.4%B				5	1○+4●	80%B	(b)
	↓									
$T = T_3 + \varDelta T$	50	35○+15●	30%B				50	15○+35●	70%B	(c)
$T = T_3 = Tp$				What is happening? (see Fig. 1.24)				↓		
$T = T_3 - \varDelta T$				80	44○+36●	45%B	20	6○+14●	70%B	(d)

Now, let us consider what happens at T_3 (see Fig. 1.32). Here, the remaining $L(\ell_3)$ reacts with the primary crystal $\beta(S_3)$ to form $\alpha(S'_3)$. Before and after the reaction the composition of β phase is the same; only its amount changes from $x'_1\ell_3$ to $x'_1s'_3$. At $T = T_3(T_p)$ A and B atoms corresponding to the quantity in shaded area (denoted by ($\beta \to \alpha$)) get out of β phase and react with L phase to form α phase, which is formed at an interface between β and L. α envelopes the primary crystal β (Fig. 1.34).

Exercise. Draw cooling curves obtained when alloys x_1 and x_2 in

Fig. 1.30(a)(b) are cooled from L very slowly. Explain reactions in relevant temperature ranges.

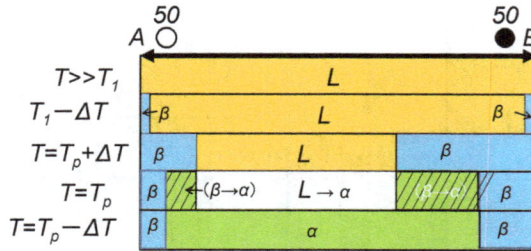

Fig. 1.32   Distribution of A and B atoms during a peritectic reaction.

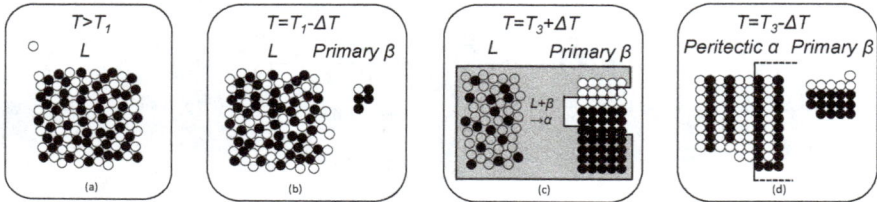

Fig. 1.33   At $T = T_3 + \triangle T$, atoms in a shadowed area react to form peritectic α phase. Out of peritectic α phase at $T = T_3 - \triangle T$ three columns at the right have moved from the primary β phase.

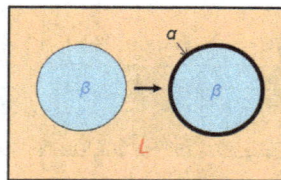

Fig. 1.34   Formation of α phase in a peritectic reaction.

1.2.7   Other peritectic-type invariant reactions

Typical peritectic reactions are summarized in Fig. 1.35.

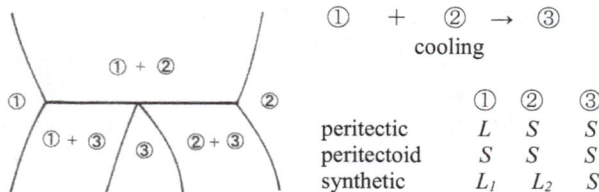

	①	②	③
peritectic	L	S	S
peritectoid	S	S	S
synthetic	L_1	L_2	S

Fig. 1.35   Peritectic type reactions.

1.2.8   Phase diagram containing an intermediate phase

In a binary A-B phase diagram, a solid solution based on A or B is referred to as a primary solid solution. By contrast, a phase which is quite different from A and B primary solid solutions can occur at a fairly high composition. Such a phase is referred to as an intermediate compound. Intermediate phases whose composition ratio A:B is a simple constant (stoichiometry) is referred to as an intermetallic compound. In practice, non-stoichiometric compounds occur frequently. The definition to distinguish between intermediate phase and intermetallic compound is not necessarily clear.

a. Intermediate phase with its own melting point

An intermediate phase $A_m B_n$ which has its own melting point is referred to as a congruent phase and is shown in Fig. 1.36. This phase diagram can be considered as a composite of two independent eutectic systems of $(A - A_m B_n)$ and $(A_m B_n - B)$ systems. That is,

$$(A - B) \text{ system} = (A - A_m B_n) \text{ system} + (A_m B_n - B) \text{ system}.$$

In other words, the intermediate phase $A_m B_n$ can be considered as one component. $(A - A_m B_n)$ and $(A_m B_n - B)$ systems are referred to as pseudo-binary systems.

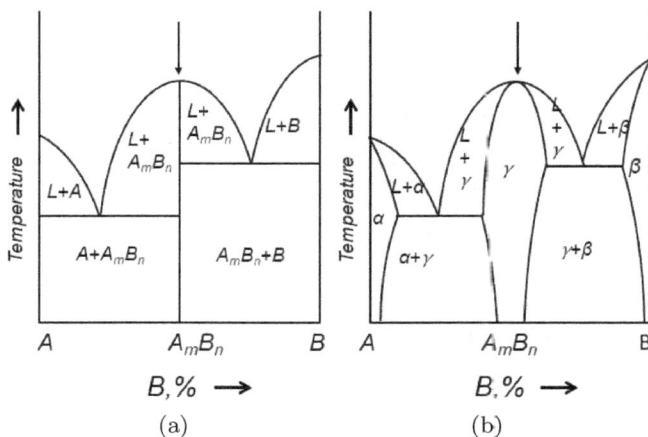

Fig. 1.36   Phase diagrams where a congruent phase $A_m B_n$ is present. (a) Solubility limits of all of $A(\alpha)$, $B(\beta)$ and the congruent phase $A_m B_n$ are negligibly small. Typical examples are Ga-As and In-Sb systems. (b) All of $A(\alpha)$, $B(\beta)$ and $A_m B_n(\gamma)$ have considerable solubility.

b. Intermediate phase without its own melting point

When $A_m B_n$ does not have its own melting point, a peritectic reaction occurs (Fig. 1.37).

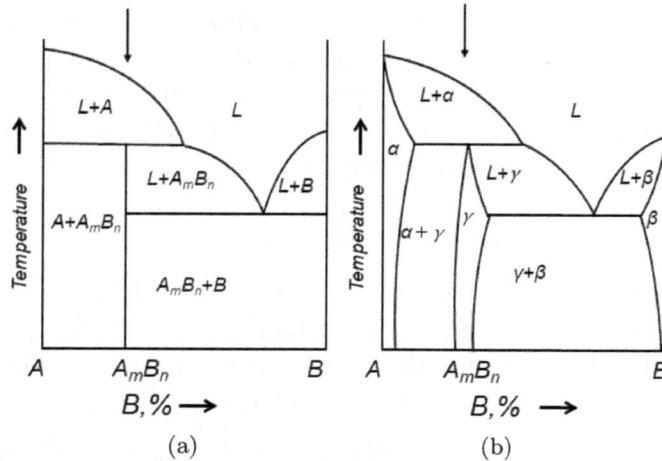

Fig. 1.37 Phase diagrams where an intermediate phase does not have its melting temperature. (a) Solubility limits of all $A(\alpha)$, $B(\beta)$ and the intermediate phase $A_m B_n$ are negligibly small. (b) All of $A(\alpha)$, $B(\beta)$ and $A_m B_n(\gamma)$ have considerable solubility.

1.2.9   Two-phase separation versus ordering

Some solid solutions (described in Sec. 1.2.2) which are completely soluble mutually at high temperatures are separated into two phases at low temperature. Figure 1.38 shows an example. Solubility limit in a solid is referred to as a solid solution curve (in short form, **solvus**). The peak temperature on the solvus line is referred to as a critical temperature (T_c).

When an alloy x is cooled from L, freezing of α-phase commences at $T_{\ell 1}$ and finishes at T_{s1}. The α-solid solution reaches the solvus at T_c and separates into α_1- and α_2-solid solutions. Here, both α_1- and α_2-solid solutions stem originally from α-solid solution, so that their crystal structures are identical. However, α_1 is richer in A, while b_1 is richer in B.

By contrast, when an alloy y is cooled from L, α_2 precipitates at T_1 from α_1-solid solution after freezing.

On the other hand, ordered alloys may be formed at low temperature. Figure 1.39 shows the phase diagram of Cu-Au system, where three types of ordered alloy (Cu_3Au, $CuAu$ and $CuAuII$) occur, the crystal structures of Cu_3Au, $CuAuI$ and $CuAuII$ being shown in Fig. 1.8(c), (b) and (e).

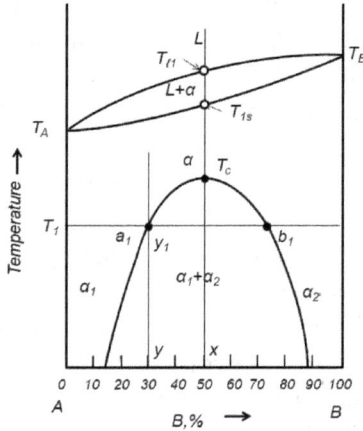

Fig. 1.38   Spinodal decomposition.

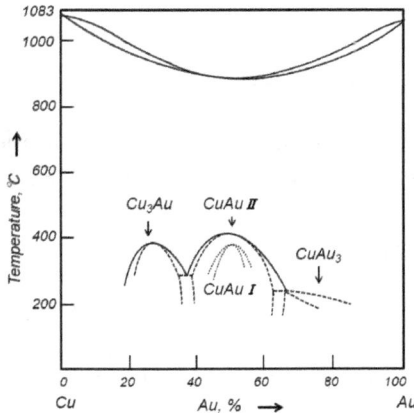

Fig. 1.39   Phase diagram of the Cu-Au binary system.

Why does an isomorphous solid solution at high temperatures separate into two phases on the one hand and transform into ordered phases on the other? This can be accounted for by considering mutual interaction between A and B atoms. A solid solution where A and B atoms behave completely in a similar manner in an A-B binary alloy is referred to as an ideal solution. In this case, A and B atoms are arranged completely at random. However, when A atoms and B atoms repulse each other, bonding of A-A and B-B will be preferred to that of A-B, resulting in clustering of A atoms and B atoms.

When clustering proceeds further, two-phase separation will take place. By contrast, when the interaction between A atoms and B atoms is attractive, bonding of A-B will be preferred to those of A-A and B-B. In other words, the neighbors of A atoms will be occupied by B atoms and vice versa, resulting in ordering. In summary both two-phase separation and ordering result from deviation from an ideal solid solution.

The reason why ordering or two-phase separation disappears at high temperatures is accounted for by the effect of entropy (S). Entropy is a factor indicating how the system under consideration is at random. The free energy of a binary A-B solid solution G is expressed by $G = H - TS$, where H is enthalpy. In this equation entropy appears in the form of $-TS$, where T is the absolute temperature, so that the higher is temperatures (with increasing temperature T) the more important is effect of entropy in lowering the free energy of the system G. As a result, a disordered solid solution is preferred (see Sec. 3.1.3 for details).

1.2.10   Summary

a. Fundamental phase diagrams

An apparently complex phase diagram can be divided into a mixture of simple phase diagrams (or reactions) as shown in Fig. 1.40(a)(b).

b. Errors in the binary phase diagram

1) Two neighboring single-phase regions (α, β) are separated by a dual phase region $(\alpha + \beta)$ (Fig. 1.41).

2) There is a total of six two-phase boundaries emanating from a three-phase isotherm. These six lines are either (4 upper + 2 lower) or (2 upper + 4 lower) (Fig. 1.42).

3) Two single-phase regions can contact each other only at one point. If the situation shown in Fig. 1.43(e) can exist, at T' a total of 2 two-phase regions, $(\alpha + \gamma)$ and $(\gamma + \beta)$, exists independently, however, at T'' they coalesence into a single two-phase region $(\alpha + \beta)$. In other words, at T, α (with composition: a), β (with composition: b) and γ (with composition: c) are coexistent. This is nothing but an invariant reaction, so that a horizontal isotherm of three-phase equilibrium connecting a, c, b is to be drawn (horizontal line in Fig. 1.43(f)). Otherwise, boundaries of two single-phase regions $(\alpha + \gamma)$ and $(\gamma + \beta)$ should contact each other either at maximum or minimum (Fig. 1.43(a) or (b)).

4) An extension of a boundary of a single-phase region (metastable phase)

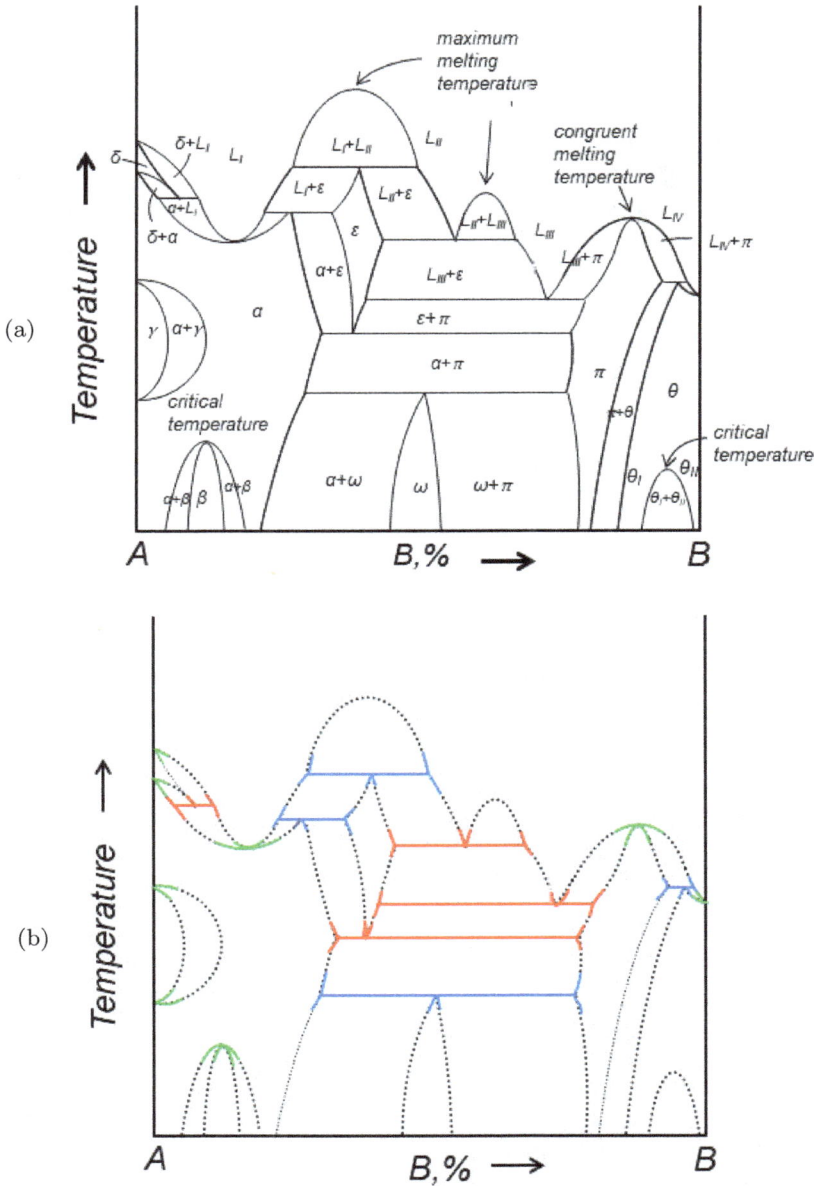

Fig. 1.40   (a) A complex phase diagram. (b) Decomposition of (a).

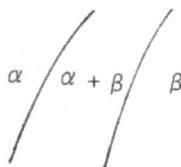

$\alpha \quad \alpha + \beta \quad \beta$

Fig. 1.41 Single phase regions (α and β) and a dual phase region ($\alpha + \beta$) in between α and β phases.

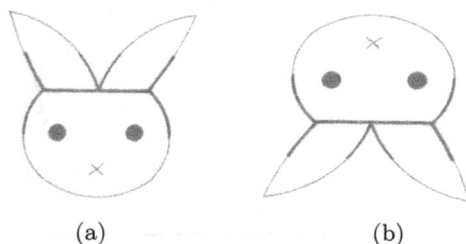

(a)   (b)

Fig. 1.42 (a) Eutectic reaction. (b) Peritectic reaction, which is the upsidedown of eutectic reaction.

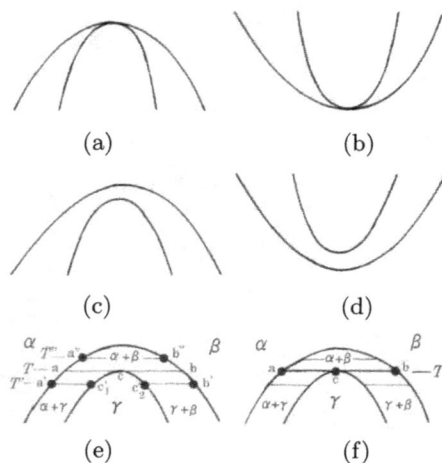

(a)   (b)

(c)   (d)

(e)   (f)

Fig. 1.43 Errors in phase diagram (I). (a) and (b) are valid, but (c) and (d) are invalid. If a horizontal line ab is added, the phase diagram is valid.

cannot exist in the original single-phase region but must exist in the adjacent two-phase region. Therefore, an angle between two boundaries of two single-phase regions is less than 180° (Fig. 1.44).

Exercise. Explain the reason.

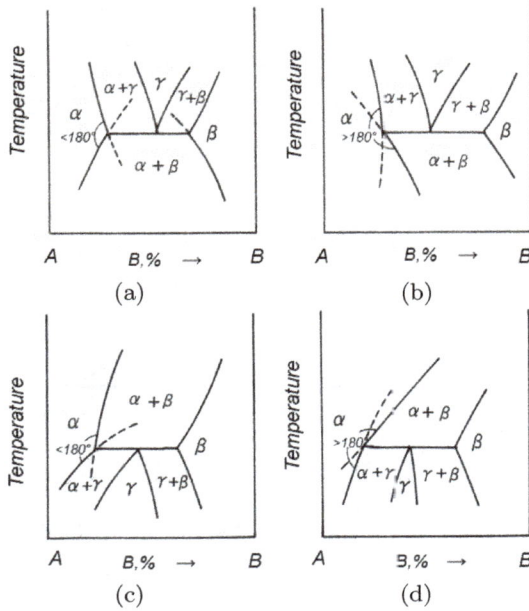

Fig. 1.44 Errors in phase diagram (II). (a) and (c) are valid, but (b) and (d) are invalid.

Hint. Let us consider the cooling process of an alloy x from dual phase region $(\alpha + L)$ in Fig. 1.45. Assume that after passing through the eutectic temperature T_e to $(T_e - \triangle T)$, which is below T_e, the alloy remains in the two-phase region of $(\alpha + L)$ without experiencing the eutectic reaction. This situation is referred to as a metastable equilibrium. The extension of the liquid is cc′ and the solidus of α solid-solution is aa′. On the other hand, at equilibrium the eutectic reaction $L \rightarrow \alpha + \beta$ takes place at T_e and L vanishes, resulting in a dual phase of $(\alpha + \beta)$.

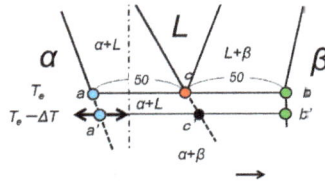

Fig. 1.45 Difference between the equilibrium concentration and the meta-stable concentration.

The composition of β is b (or b′ very close to b), which is much B-richer than the metastable composition c′. In order for this to occur, should the composition of α be deviated from a′ toward B-rich or A-rich?

5) Other erroneous examples

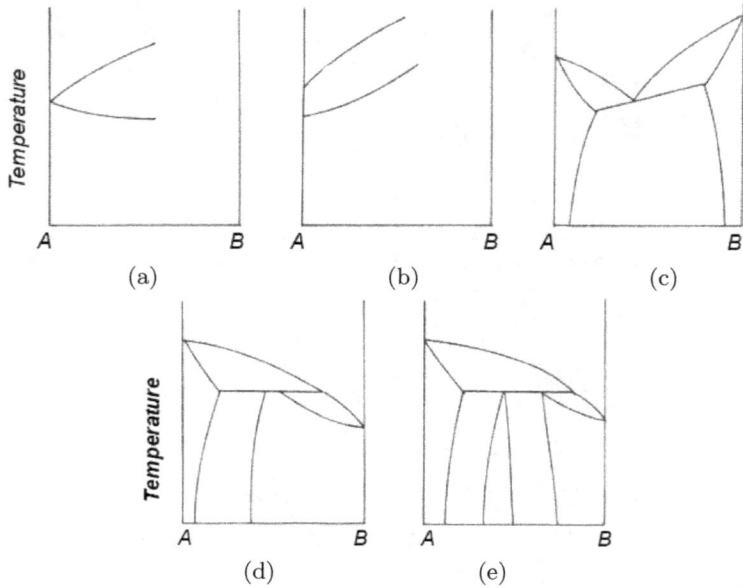

Fig. 1.46   Errors in phase diagram (III).

Exercise. Consider the reasons why Fig. 1.46 is erroneous.

Exercise. Suppose that B element is added to a pure metal of A element to form α solid-solution and then the lattice constant of the α solid-solution is measured as a function of the content of B at T. Let us assume that the atomic radius of B element is larger than that of A element. In this case, which of ABC, ABD, ABE in Fig. 1.47 is correct?

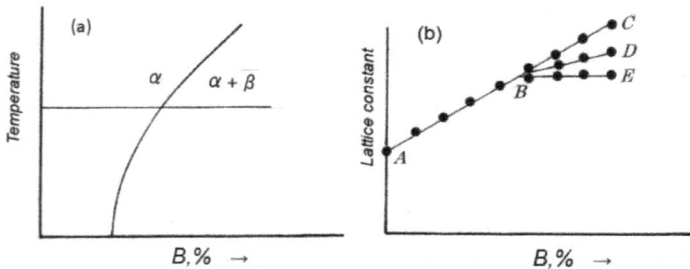

Fig. 1.47   Concentration dependence of lattice constant.

Exercise. There are errors in Fig. 1.48, where the broken line indicates a magnetic transformation. Draw the right phase diagram.

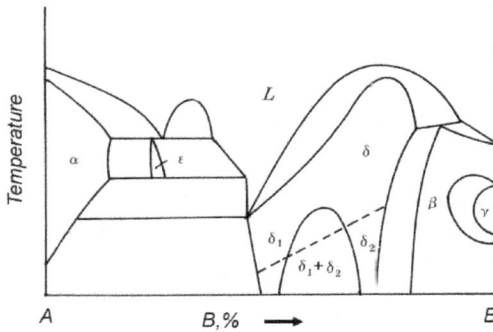

Fig. 1.48 Correct errors in the phase diagram.

Exercise. Figure 1.49 is unfinished. Finish this.

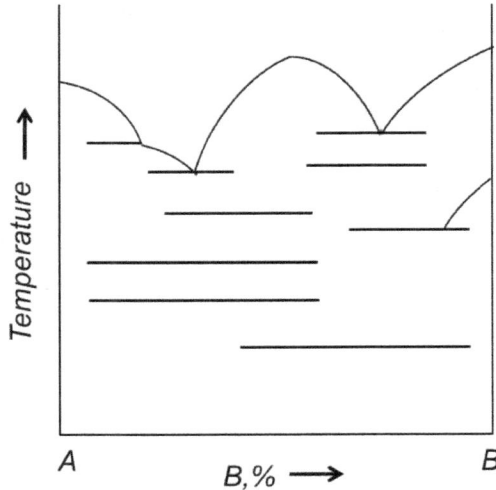

Fig. 1.49 Complete the phase diagram.

Exercise. Constitute the binary Hf-Th phase diagram between 1000°C–2400°C under the following conditions. Composition is denoted in at.% and put names of phases in the respective regions.

(I) Melting point of Hf: 2230°C, melting point of Th: 1760°C.

(II) In Hf a phase transformation takes place at 1750°C. The low temperature phase is α and the high temperature phase is β. Similarly, in Th a phase transformation takes place at 1360°C. The low temperature phase is γ and the high temperature phase is δ.

(III) The α solid-solution contains a small amount of Th. The solubility limit is 2%Th in a temperature range of 1300°C–1600°C and 1% at 1000°C.

Similarly, β solid solution contains a small amount of Th. The solubility limit of β is 5%Th at 1600°C.

The γ solid-solution contains Hf. The solubility limit of the γ solid solution is 5%Hf at 1300°C and 3%Hf at 1000°C. Similarly, δ solid solution contains Hf and the solubility limit is 17%Hf at 1450°C.

(IV) In this alloy system three invariant reactions are present as follows, where the phases in the left-hand side indicate high-temperature phases.

   a) At 1600°C: β solid-solution (5%Th)$\rightarrow$ α solid-solution (2%Th) + liquid (57%Th)

   b) At 1450°C: Liquid (68%Th)$\rightarrow$ α solid-solution (2%Th) + δ solid-solution (83%Th)

   c) At 1300°C: δ solid-solution (90%Th)$\rightarrow$ α solid-solution (2%Th) + γ solid-solution (95%Th)

(V) Approximate all the phase boundaries by straight lines.

1.3 Deviation from Equilibrium in Natural Cooling of Binary Alloys

1.3.1 Isomorphous (complete) solid solution

Both liquid and solid phases are involved in freezing. Diffusion in a solid is much slower than that in a liquid. Thus, in natural cooling deviation from equilibrium takes place, resulting in a microstructure which is quite different from those obtained in a slow (equilibrium) cooling.

Here, the following will be assumed:

1) Diffusion in a solid is much slower than that in a liquid.
2) Diffusion in a liquid is rapid enough.
3) The solid-liquid interface is at equilibrium.

Let us consider processes which take place when an alloy liquid with composition ℓ_1 in Fig. 1.50 is solidified. At T_1 a solid with composition s_1 is crystallized from a liquid with ℓ_1. On further cooling down to T_2 slowly enough to keep equilibrium, diffusion takes place even in a solid, resulting in the formation of a totally homogeneous composition of s_2. In natural cooling a solid of s_1 crystallizes at T_1, which is covered by a newly solidified solid

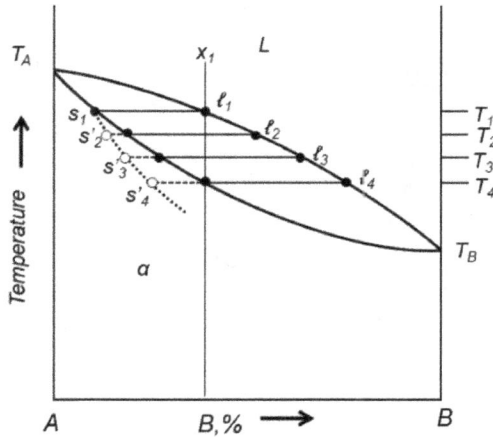

Fig. 1.50 Solidus in natural cooling (broken line).

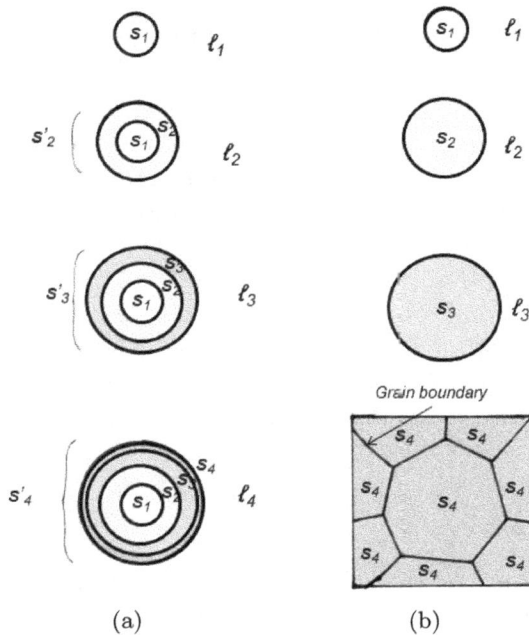

(a)   (b)

Fig. 1.51 Solidified structures in (a) natural cooling and (b) equilibrium cooling.

s_2. As a result, the composition inside a solid differs from surface (s_2) to interior (s_1), the average composition being s_2' somewhere between s_1 and s_2 (Fig. 1.50). The structure in which the composition differs from center to periphery is referred to as a cored structure. The phenomenon where

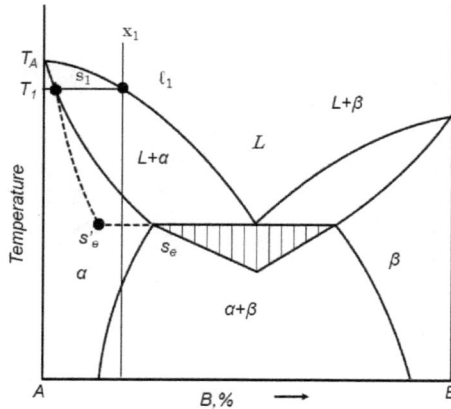

Fig. 1.52 Extension of eutectic reaction in natural cooling. In equilibrium cooling, in an alloy whose composition of β is lower than s_e, eutectic reaction does not occur while in natural cooling eutectic reaction occurs upto s'_e.

solute atoms are distributed unevenly and present in excess is referred to as segregation.

Thus, in natural cooling B is deficient than at equilibrium, lowering the solidus line to lower temperature. At equilibrium an alloy X_1 is solidified completely at T_4 with composition s_4, however, in natural cooling solidification is unfinished with a liquid of composition ℓ_4 remaining.

1.3.2 Eutectic alloy

In a eutectic alloy the solidus is shifted downward in cooling. As a result, even in an alloy for whose composition a eutectic reaction does not occur in the equilibrium cooling (for instance x_1) a eutectic reaction occurs (Fig. 1.52).

1.3.3 Peritectic alloy

For peritectic alloys, the deviation from equilibrium in natural cooling is quite large. The reason for this is that product (α) of a peritectic reaction $L + \beta \rightarrow \alpha$ is formed at an interface between L and β. Once α phase is formed between β and L, in order for a reaction between A and B atoms to continue, A and B atoms must cross α phase as shown in Fig. 1.53. This retards the growth of α phase significantly and often prevents its completion.

Let us suppose that an alloy of x in Fig. 1.54(a) is cooled from liquid naturally,

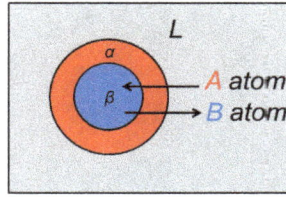

Fig. 1.53 Migration of A and B atoms in peritectic reaction.

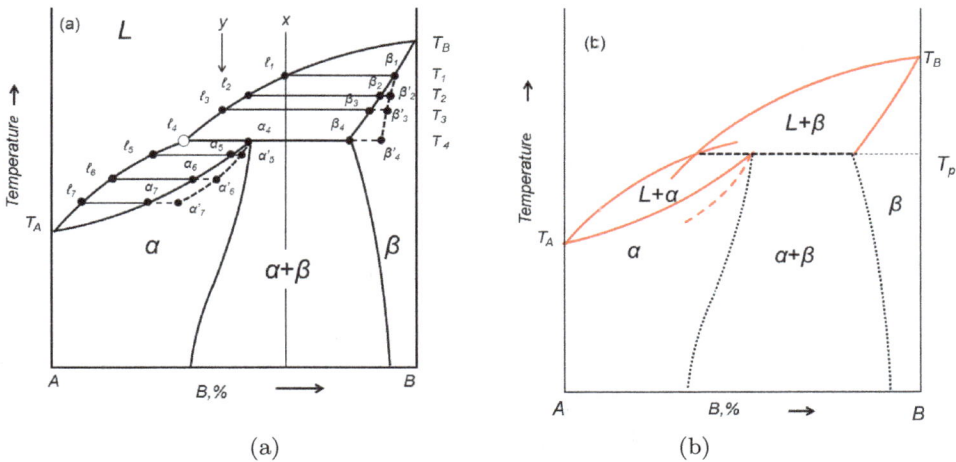

(a)   (b)

Fig. 1.54 (a) Equilibrium cooling in peritectic reaction. (b) In natural cooling peritectic reaction is unlikely to complete.

① $T_4 + \Delta T < T < T_1$ A primary β phase is crystallized in a cored structure. Average composition of β changes in the order of $\beta'_1 \rightarrow \beta'_2 \rightarrow \beta'_3 \rightarrow \beta'_4$.

② $T = T_4 + \Delta T$

$$[\%\beta'_4] = x_4\ell_4/\beta'_4\ell_4 < x_4\ell_4/\beta_4\ell_4$$
$$[\%\ell_4] = \beta'_4\, x_4/\beta'_4\ell_4 > x_4\beta_4/\beta_4\ell_4$$

That is, β'_4 is richer in β than at equilibrium.

③ $T = T_4$ Reaction $\ell_4 + \beta'_4 \rightarrow \alpha_4$ proceeds partially with α phase formed at an interface between L and β phases. The subsequent peritictic reaction proceeds only when A and/or B atoms diffuse across α phase. α phase is a solid: The diffusion in a solid is quite sluggish, so that the rate of the reaction is quite slow and practically the reaction stops.

④ $T < T_4$ α phase crystallizes directly from the remaining ℓ_4 as if it were the primary phase. As a result α phase envelopes the pre-existing nucleus

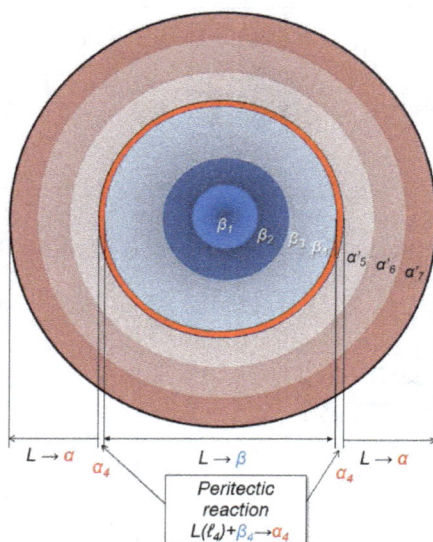

Fig. 1.55 Formation of a cored structure in natural cooling of a peritectic alloy.

of β phase (the average composition of α changing from $\alpha_5' \to \alpha'$). The resultant structure is a cored structure with α phase surrounding a nucleus of β phase (Fig. 1.55).

⑤ $T = T_7$ The remaining L_4 $(\ell_4 - \alpha_7')$ transforms to α solid solution with α_7 after passing its solidification period.

Exercise. Discuss reactions which happen in an alloy of y in natural cooling.

Thus, in a peritectic alloy α phase is formed temporarily by peritectic reaction $L + \beta \to \alpha$ at the peritectic temperature, but thereafter only solidification of $L \to \alpha$ proceeds as if the solvus indicating two-phase region $(\alpha + \beta)$ does not exist (Fig. 1.54(b)). In other words, it is quite difficult to grow to an intermediate phase which shows a peritiectic reaction from a liquid.

Chapter 2

Heat Treatment of Alloys

Phase diagrams treat only equilibrium states. As discussed in Chapter 1, however, the equilibrium states are not always attained in practical processes. However, this does not necessarily mean that phase diagrams are of no (or little) use in interpreting non-equilibrium phenomena.

The definition of equilibrium states is illustrated in Fig. 2.1, where the horizontal line shows the reaction path and the vertical line shows free energy. Unstable equilibrium corresponds to the summit. If you stumble one step, you drop into the valley, which corresponds to stable equilibrium. This is the reason why the summit is referred to as an unstable equilibrium. By contrast, in the valley even if you step out of the way, you will return to the valley. In this sense it is a stable equilibrium. Suppose that there are many valleys. The state having the lowest energy corresponds to the genuine stable equilibrium; other states having higher energy are referred to as meta-stable

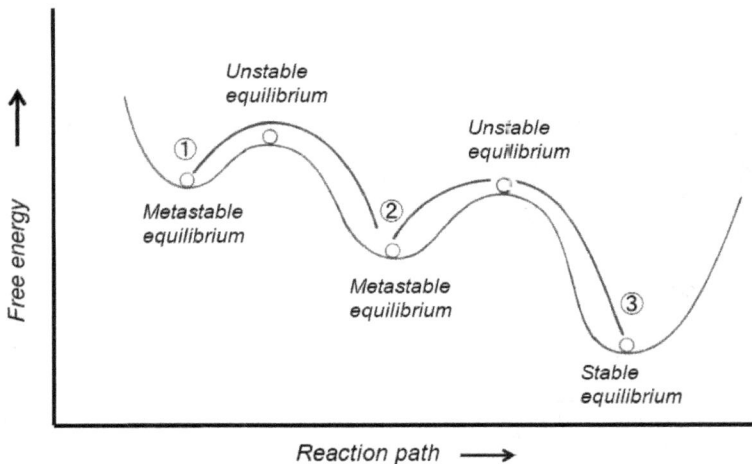

Fig. 2.1 Meta-stable equilibrium, unstable equilibrium and stable equilibrium.

equilibrium. A meta-stable equilibrium is eventually doomed to change into the genuine stable equilibrium. However, the time to reach the genuine stable equilibrium depends on individual reactions. Human beings have taken most of the deviation from the equilibrium state skillfully to achieve superior properties of alloys. This is called heat treatment. In this part the principle of heat treatment will be described from the viewpoint of phase diagrams.

2.1   Aging and Precipitation in Aluminum Alloys

2.1.1   Explanation based on phase diagram

In general, a solubility limit in Al alloys is quite small, but definitely non-zero.

Figure 2.2 shows the Al side of the phase diagram of binary Al-Cu system. A solid solution based on pure Al (α_{Al}) is referred to as a primary solid solution (see Sec. 1.1.2g). Next to α_{Al} an intermediate phase θ (CuAl$_2$) is present (see Sec. 1.2.8). The solubility limit of Cu in α_{Al} phase is 5.7 wt.% at 548°C, but decreases with decreasing temperature, becoming negligibly small at room temperature. Suppose that an Al-4wt.%Cu alloy is heated at 548°C for a long enough time for Cu atoms to dissolve completely in Al, to make a complete solid solution (this treatment is referred to as solution treatment,

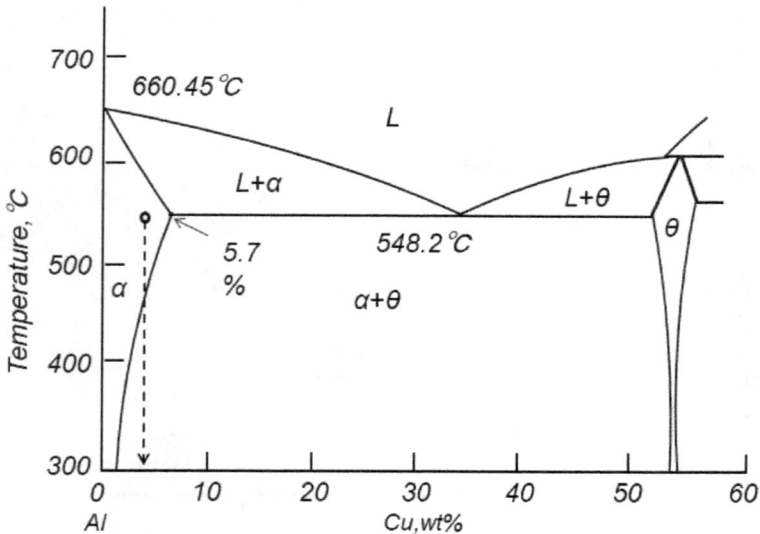

Fig. 2.2   Al side of the binary Al-Cu phase diagram. (θ is Al$_2$Cu.)

ST), followed by slow cooling, for instance, in furnace cooling (FC), where the power of an electric furnace is killed while keeping the alloy inside the furnace. Then, at around 470°C, a reaction $\alpha \to \alpha_{Al} + \theta$ takes place and θ phase whose crystal structure is different from that of α_{Al} emerges. This phenomenon is referred to as precipitation (by contrast a reaction where a crystal emerges from a liquid is referred to as freezing, solidification or crystallization). θ phase is called a precipitate, and α_{Al} a matrix.

Naturally, in order for precipitation to take place in reality, Cu atoms must travel a considerable distance in α_{Al} and condense to a certain concentration to form θ phase somewhere. This movement is referred to as diffusion.

However, when the alloy is cooled rapidly by, say, water quench (WQ) or oil quench (OQ) or, in some cases, even air cooling (AC), the alloy is cooled down to room temperature without diffusion of Cu taking place. Diffusion is rapid at high temperature but sluggish at room temperature. In other words, α_{Al} solid solution is enforced to contain Cu atoms, which at equilibrium cannot be present inside, and brought to room temperature. Such a solid solution is referred to as a supersaturated solid solution.

Now let us consider the situation in a supersaturated solid solution. Matrix α_{Al} is not able to accommodate Cu atoms, and Cu atoms want to exit from the supersaturated solid solution of α_{Al} to form independent θ phase. The diffusion rate of Cu atoms is low at room temperature but not zero. As the time lapses Cu atoms diffuse and coagulate and eventually precipitation takes place. This phenomenon is referred to as aging. Precipitates strengthen the alloy. This is referred to as precipitation hardening, or age hardening. When aging takes place at room temperature it is called natural aging. If a supersaturated solid solution is aged at a higher temperature, say 100°C, the diffusion of Cu atoms is enhanced, resulting in acceleration of precipitation. Such an aging is referred to as artificial aging.

2.1.2 Morphology of precipitates, structure of interfaces

Precipitates can assume a variety of morphology including sphere, plane and needle (Fig. 2.3). The morphology is controlled by the following factors:

① Interfacial energy between the matrix and the precipitates,
② Strain energy associated with precipitation,
③ Orientation dependence of crystal growth.

Among them, ① and ② stem from increases in energy associated with precipitation. When interfacial energy and strain energy are not dependent

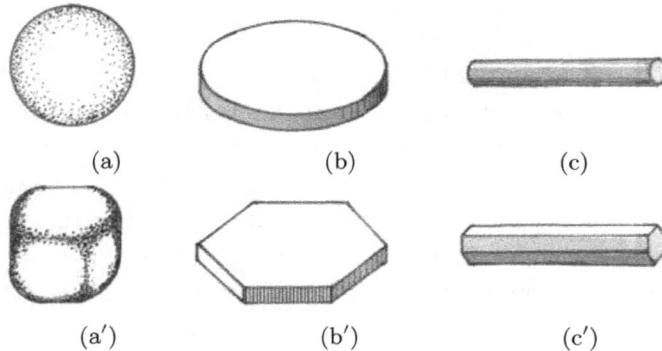

Fig. 2.3 Morphology of precipitates. (a) Sphere. (a′) Cuboid. (b), (b′) Plate. (c), (c′) Needle.

on orientation, a precipitate assumes spherical shape in order to minimize the interfacial area. This can be visualized by a water droplet.[1] When the energies are orientation dependent, planes with minimum energy are preferred.

On the other hand ③ is a kinetic problem and planes with minimum growth rate appear eventually.

When structures of a precipitate and the matrix are similar, a mismatch at the interface is small. Therefore, when the precipitate is small, lattices between the matrix and a precipitate can establish a one-to-one correspondence. Such an interface is referred to as a coherent or epitaxial interface. However, on increasing the size of a precipitate, a mismatch between the precipitate and the matrix is accumulated. When the mismatch reaches a spacing between the lattice points, an interfacial dislocation is formed. This state is referred to as semi-coherent. When there is no correspondence in lattice between a precipitate and the matrix, it is referred to as non-coherency. When some interfacial planes are coherent while other planes are incoherent, it is referred to as partial coherent. It is important not to confuse semi-coherency with partial coherency.

---

[1]The micrographs show a small particle of Sn in crystalline state (a) and liquid state (b). In a crystal (a) surface energy is orientation dependent, so that the particle assumes a polygon. In a liquid (b) surface energy is orientation independent, so that the particle assumes a perfect sphere.

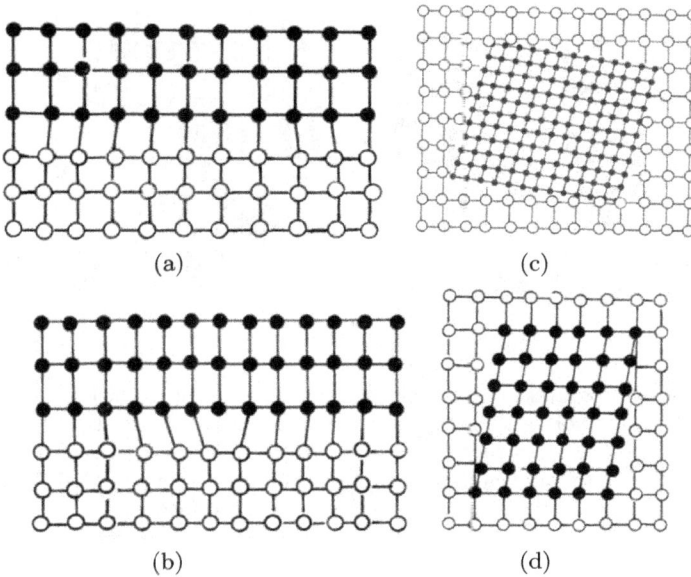

Fig. 2.4 Structure of an interface between a precipitate and the matrix. ● is a precipitate and ○ is the matrix. (a) Coherent interface. (b) Semi-coherent interface (dislocations are introduced in the interface). (c) Incoherent interface. (d) Partial coherent interface (the horizontal interface is coherent but the vertical one is incoherent).

When the crystal structure of a precipitate is quite similar to that of the matrix and their lattice constants are very close, a precipitate can stay coherent after growth. A typical example for such an interface is the primary solid solution (γ) and $Ni_3Al(\gamma')$ in the Ni-Al system. Ni crystallizes in FCC with $a = 3.5238$ Å, and γ' crystallizes in $L1_2$, which is based on FCC with $a = 3.566$ Å: They are very similar. Under such a circumstance, coherency is maintained after precipitates become considerably large. Figure 2.5(a) shows a SEM and (b) shows HREM of the interface of γ/γ'.

Tip

Usually, most materials weaken with increasing temperature. However, γ' is exceptional in that it hardens on increasing temperature. This exceptional property is referred to as inverse temperature dependence of strength (Fig. 2.6). Super heat-resisting alloys, consisting of a mixture of γ/γ', make use of this property.

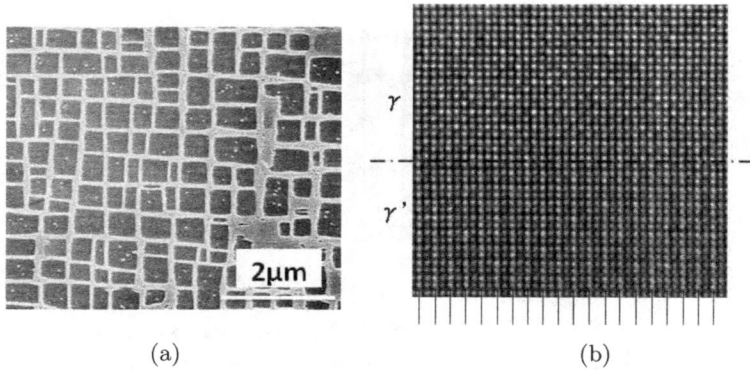

(a)   (b)

Fig. 2.5 SEM photograph of γ' (black region) and γ (grey region) phases in a Ni-base superalloy. The alloy consists mostly of γ' (courtesy of Dr. Murata). (b) HREM of a γ/γ' interface. γ has FCC structure while γ' has $L1_2$ (see Fig. 1.9(b)), the spacing of lattices which is twice as large as that of γ. Vertical lines drawn below the picture indicate superlattices of γ' (lattice constant is 0.357 nm). The interface is perfectly coherent.

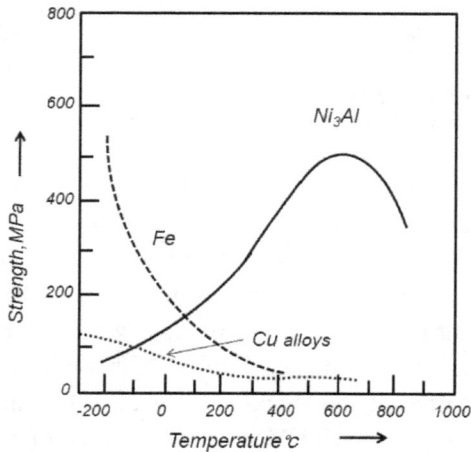

Fig. 2.6 Temperature dependence of strength (schematic). Ni_3Al shows the inverse temperature dependence of strength.

2.1.3   Sequence of precipitation, G.P. zone and metastable precipitates

The foregoing discussion is what the phase diagram says. However, in reality θ phase rarely precipitates directly from a supersaturated α_{Al} solid solution. This is especially true when aging temperature is low. Aging phenomenon can be compared to travelling from the starting station (for example, Tokyo) to the terminal station (for example, Osaka).

When aging temperature and hence the reaction rate is high, you travel from Tokyo directly to Osaka by air plane without calling at any other places. This corresponds to travelling from metastable state ① to stable equilibrium ③ in Fig. 2.1. If the temperature is moderate, you travel by Shinkansen (bullet train), calling at Nagoya and Kyoto.[2] If the diffusion rate is still lower, you travel via the old railway. Depending on what ways you travel, the numbers of calling-at stations and the route itself are different. The station where you call at is a semi-stable precipitate.

Figure 2.7 shows the Al-Cu phase diagram with solvus for *metastable* G.P. zone and θ''. In this sense it is not the equilibrium phase diagram, however, for the sake of convenience this type of *phase diagram* is often used.

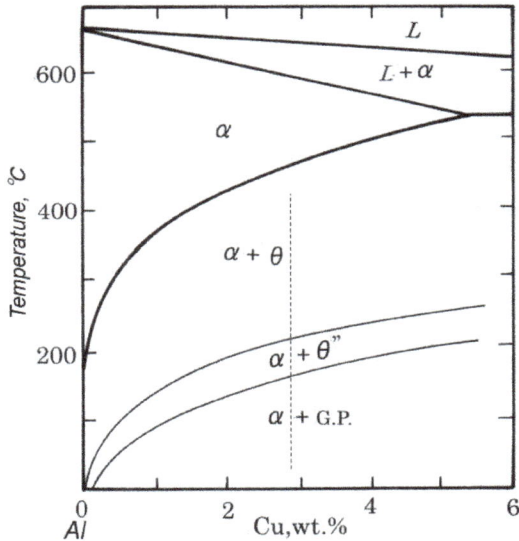

Fig. 2.7 Phase diagram showing occurrence of intermediate phases such as G.P. zone (Al-Cu). (Reproduced from Beton, R. H. and Rollason, E. C., *J. Inst. Metals*, 86 (1957–58) 77.)

---

[2]Map of central Japan.

(a)   (b)

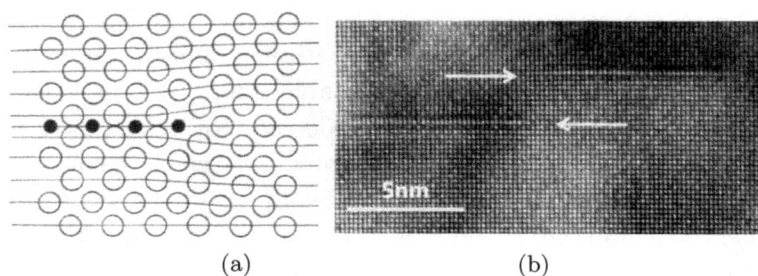

Fig. 2.8 (a) Schematic illustration of G.P.(I) zone in an Al-Cu alloy. ○ stands for Al atoms of the matrix and ● Cu atoms. Cu atoms lie on a {001} plane of an Al matrix. (Reproduced from Gerold, V.: *Z. Metallk*, 45(1954)593, 599.) (b) HREM of G.P.(I) zone. Horizontal atomic rows appearing whiter (indicated by arrows) are G.P. zone (courtesy of Dr. Y. Konno).

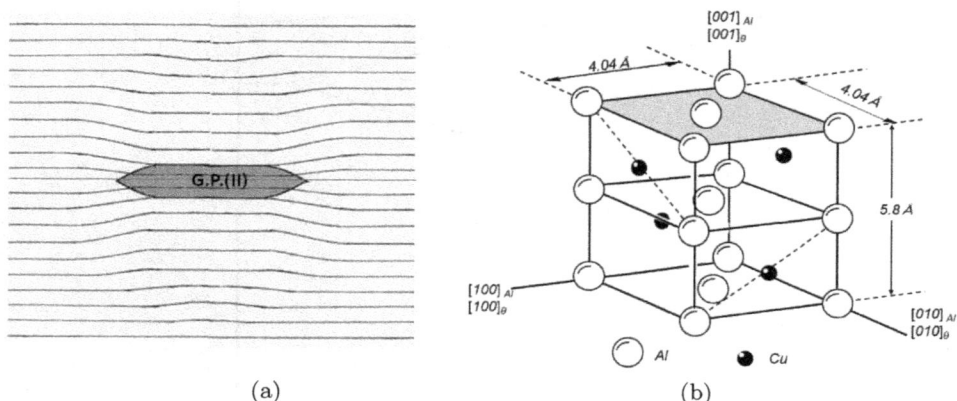

(a)   (b)

Fig. 2.9 (a) G.P.(II). G.P.(II) is occasionally denoted by θ''. Horizontal (or nearly horizontal) lines indicate the (001) planes of Al matrix. (b) θ' phase.

When the aging temperature is low (130°C), precipitation proceeds in the order of

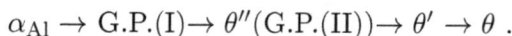

$$\alpha_{Al} \rightarrow \text{G.P.(I)} \rightarrow \theta''(\text{G.P.(II)}) \rightarrow \theta' \rightarrow \theta .$$

This is referred to as precipitation sequence. Here, G.P. means Guinier-Preston zone which was named after the discoverers Guinier[3] and Preston.[4] In G.P.(I) one atomic thick Cu atoms precipitate on the Al matrix {001} (Figs. 2.8(a) and (b)). In G.P.(II) the thickness of Cu increased (Fig. 2.9(a)). G.P.(I) and G.P.(II) are coherent with the matrix. In θ' (001) plane (hatched in Fig. 2.9(b) $a_{\theta'} = 4.04$ Å) is coherent with (001) plane of Al ($a_{Al} = 4.049$ Å), however, the mismatch for (100) and (010) planes is too large

---

[3] André Guinier (1911–2000), crystallographer, France.
[4] George Dawson Preston (1896–1972), physicist, UK.

for coherency. In other words, θ' is a metastable precipitate with partial coherency. The equilibrium θ phase is perfectly incoherent with the matrix.

When the aging temperature is a bit higher such as 200°C, the precipitation sequence becomes

$$\alpha_{Al} \to \theta' \to \theta$$

skipping G.P. zone. G.P. zone was first named after Guinier and Preston for the metastable phases in Al-Cu system. However, later this naming has been applied to metastable phases in general which are formed at early stage of precipitation. Thus, the shape of the G.P. zone is not necessarily planar. For instance G.P. zone in Al-Ag system is spherical.

Exercise. Show that θ' has composition of $CuAl_2$ from Fig. 2.9(b).

2.1.4 Reversion

When an Al-3wt.%Cu alloy is aged at 100°C (see Fig. 2.7), G.P. zones precipitate. Suppose that then the alloy is heated at 300°C for a short time. The G.P. zones disappear (dissolved) instantaneously because of their small size. On the other hand, it is too early for the equilibrium θ phase to precipitate. In other words, the alloy returns to the initial supersaturated α-phase. At the same time, hardening which resulted from aging at 100°C disappears. This phenomenon is referred to as reversion.[5]

2.1.5 Precipitation hardening

Precipitation brings out hardening of an alloy. This is referred to as precipitation hardening. The motion of dislocations which are responsible for plastic deformation is hindered by precipitates.

Figure 2.10 shows variation in hardness in Al-Cu alloys during aging at 130°C. This curve is referred to as an aging curve. For the meaning of Hv refer to Fig. 2.35.

Dislocations experience resistive force from solute atoms in a solid solution and their motions are hindered as compared with the motion in pure metals. The motion of dislocations in pure metals can be compared to driving a car on a well-paved road. The motion in a solid solution can be compared to driving a car on a damp road experiencing resistance by solid

---

[5]Metals harden when deformed plastically. This is referred to as work hardening (see Fig. 2.29 for definition of plastic deformation). When work-hardened metals are heated at high temperatures, they soften. The early stage of softening is referred to as recovery. Do not confuse reversion with recovery.

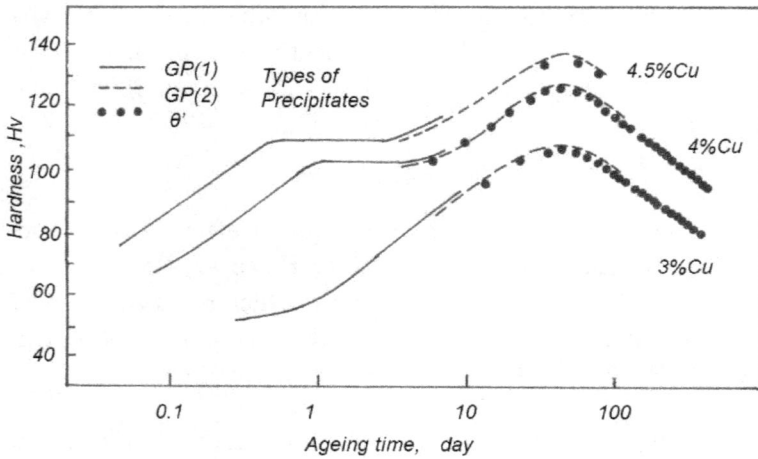

Fig. 2.10 Age-hardening curves of Al-Cu alloys at 130°C. (Reproduced from J. M. Silcock, T. J. Heal and H. K. Hardy, *J. Inst. Metals*, 82(1953–54)239: also from Hardy and Heal, *Progress in Metal Physics* 5, Fig. 63, p. 195.)

atoms, leading to hardening. This phenomenon is referred to as solution hardening.

When precipitates are very fine, resistance gets stronger with increasing size of the precipitates. As the precipitates grow further, dislocations avoid the precipitate so that the total length of the dislocation becomes longer. Since the energy of a dislocation is proportional to the total length, the total energy of the dislocation (system) increases (Fig. 2.11), resulting in hardening (Mott–Nabarro's[6] theory).

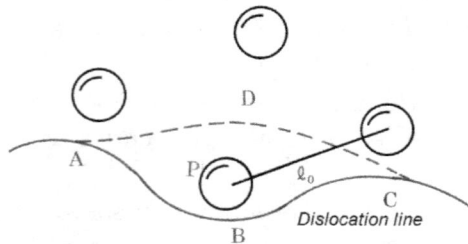

Fig. 2.11 Mechanism of precipitation hardening. Precipitates are distributed with an average interspacing of ℓ_0. A dislocation line A-B-C, which contributes to deformation, is moving upward but is blocked by a precipitate P. In order for the dislocation to move further, it must overcome P.

---

[6]Neville Mott (1916–1996), UK. Frank Nabarro (1916–2006), South Africa.

When the precipitate grows further, the spacing between them also increases. Thus, dislocation can surround the precipitate completely (Fig. 2.12(c)). As a result, dislocations can bypass precipitates, leaving dislocation loops behind. This mechanism is referred to as Orowan's[7] bypass mechanism. At this stage precipitates are not strong barriers to the motion of dislocation any longer, and the alloy softens. This phenomenon is referred to as overaging softening.

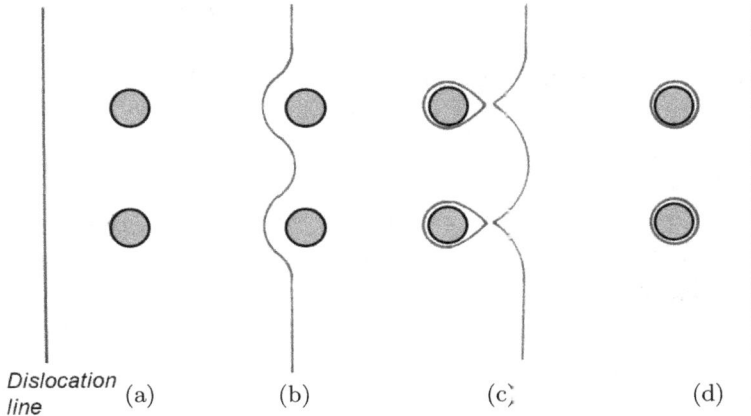

Fig. 2.12 Orowan's bypass mechanism. A dislocation line moving from left to right is blocked by precipitates. When the interspacing of the precipitates is large, the dislocation surmounts the precipitates, leaving dislocation loops (Orowan loops) behind.

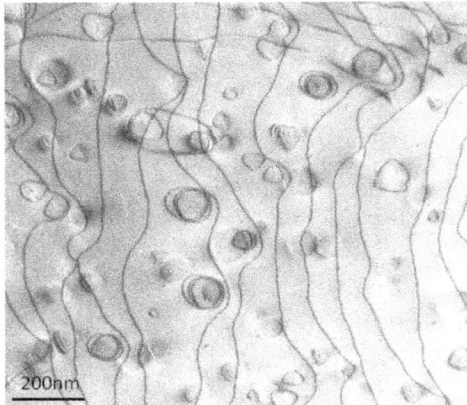

Fig. 2.13 TEM of Orowan loops (reproduced with permission from Humphrey, J. F.: *Understanding Materials*, A Festschrift for Sir Peter Hirsch, edited by C. J. Humphreys (Maney Publishing, 2002)).

---

[7]Egon Orowan (1902–1989), Hungary.

2.1.6   Spinodal decomposition

In the foregoing discussion, concentration of the solute atoms in a super-saturated solid solution is not high. For instance, concentration of Cu in an Al-5wt.% Cu alloy shown in Fig. 2.2 is ~2at.% Cu. On the other hand, concentration of Cu in the precipitate θ phase is ~33at.%. In other words, in order for θ phase to precipitate Cu atoms must be condensed to more than 15 times that in the supersaturated solid solution. This is the reason why in the early stage of precipitation G.P. zone precedes the final equilibrium θ phase.

Even so, it is a prerequisite that Cu atoms must be condensed considerably. This cannot be achieved by thermal fluctuation resulting from thermal vibration of atoms, but Cu atoms must travel a long distance (i.e., diffusion) to meet with one another. This suggests that precipitation does not start immediately after aging starts but there is a time lag. The time lag before precipitation starts is referred to as an incubation period (Fig. 2.14 ①). During the incubation period, through both thermal fluctuation and diffusion, a nucleus as the smallest precipitate is formed. Pre-nucleus zone is referred to as an embryo. Embryo is a precursor of a nucleus.

On the other hand, suppose that, in a phase diagram where two-phase separation takes place as shown in Fig. 1.38, an alloy with composition corresponding to critical temperature T_c ($x = 50$at.%B) is cooled rapidly (quenched) from just above T_c, followed by aging at T_1 in the ($\alpha_1 + \alpha_2$) two-phase region.

When quenched, α solid solution has a disordered structure, so that both A and B atoms are distributed statistically at random (Fig. 2.15(a)). However, in such a concentrated alloy as 50at.%B, with respect to atomic arrangement in a single atom of, say, B atom, there is a statistical probability of B atoms occupying the nearest neighbors. That there is a two-phase region of ($\alpha_1 + \alpha_2$) in Fig. 1.34 stems from the fact that A atom wants to have A atoms as the nearest neighbors and that B atom wants to have B atoms as the nearest neighbors. In other words, in α_1 phase A–A bond is preferred (happier) and in α_2 phase B–B bond is preferred, leading to clustering of A atoms and B atoms. Thermodynamically the total free energy of the system (alloy) is lowered, or equivalently chemical potentials of both A and B atoms are lowered (Fig. 2.15(b)). The extremely opposite of this situation is ordering where A–B bond is preferred (Fig. 2.15(c)).

Now let us consider aging of ~50at.%B alloy(x_1) at T_1 (Fig. 2.16). Before aging ($t = 0$), average concentrations of A and B atoms are represented by the horizontal line ($t = 0$). However, as is obvious from Fig. 2.15, A-rich

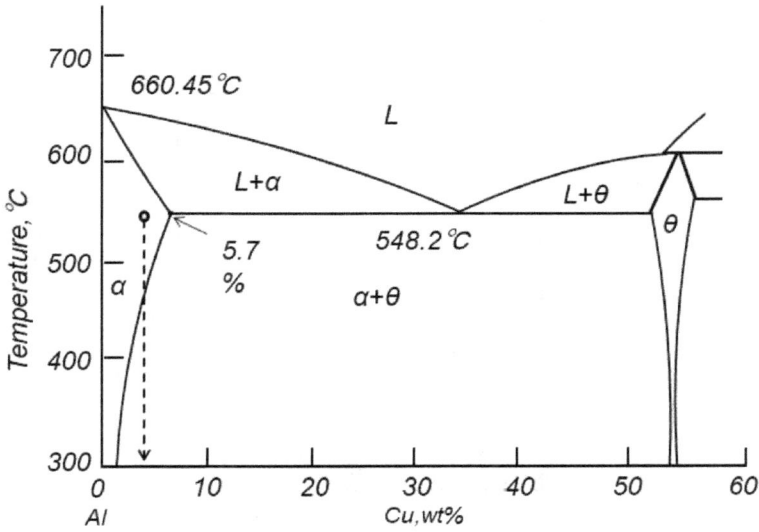

Fig. 2.2 (repeat).

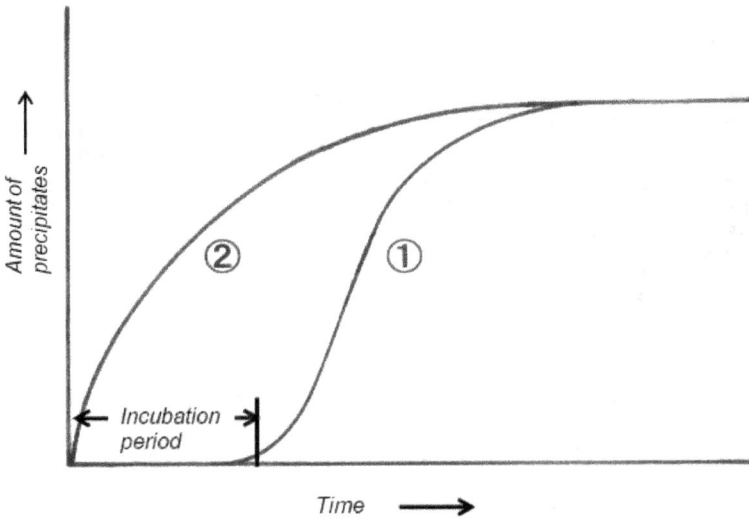

Fig. 2.14 Precipitation process with (①) and without (②) incubation period.

and B-rich zones exist inevitably. By thermal fluctuation B atom in an A-rich zone is replaced by A atom, the concentration of A atom in the A-rich zone increases, leading to an equilibrium. In other words, reaction of $\alpha \to (\alpha_1 + \alpha_2)$ proceeds without barrier. Therefore, an incubation period for nucleation is not required and reaction proceeds as indicated by ② in Fig. 2.14.

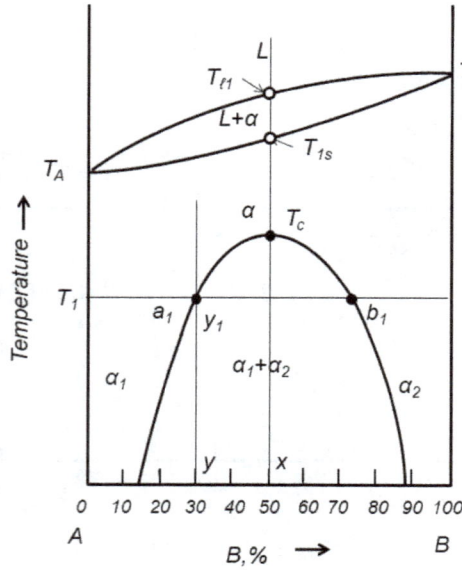

Fig. 1.38 (repeat).

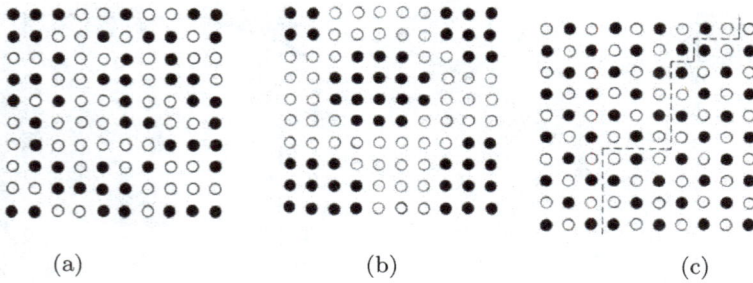

(a)   (b)   (c)

Fig. 2.15   Atomic arrangements in a solid solution. (a) Disorder. (b) Order. (c) Cluster.

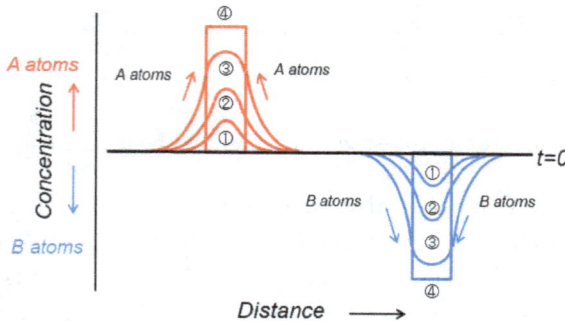

Fig. 2.16   Variation in composition in spinodal decomposition. ① $t = t_1$, ② $t = t_2$, ③ $t = t_3$ and ④ $t = t_4$ (equilibrium state).

On the other hand, a reaction where α_2 precipitates from a supersaturated solid solution with composition y_1 (say, 30%B) is fundamentally the same as a nucleation and growth transformation. To summarize, in the central zone in ($\alpha_1 + \alpha_2$) two-phase separation zone, incubation-free precipitation takes place (corresponding to ② in Fig. 2.14), while in near-solvus region an incubation-accompanying reaction takes place (corresponding to ① in Fig. 2.14). The former is referred to as spinodal decomposition and the latter as binodal decomposition. The boundary between spinodal and binodal decomposition deduced by thermodynamic consideration is shown in Fig. 2.17.

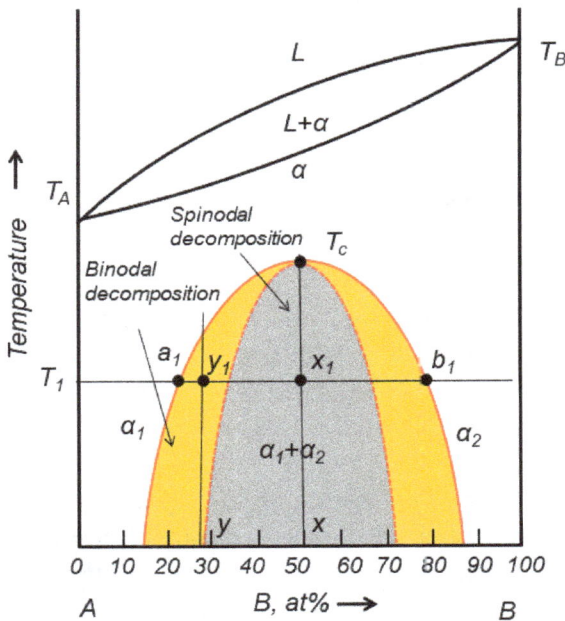

Fig. 2.17 Spinodal decomposition.

A microstructure of the spinodal decomposition is quite fine. An example is shown in Fig. 2.18. In such a microstructure composition and/or lattice parameter vary periodically, and is referred to as a modulated structure.[8]

---

[8]In spinodal decomposition, a fine modulated structure occurs. However, occurrence of a modulated structure does not necessarily indicate occurrence of spinodal decomposition since even in precipitation process via nucleation a similar structure can emerge especially in its early stage due to coherency with the matrix etc. Occurrence of a modulated structure is a necessary condition and not a satisfactory one for spinodal decomposition.

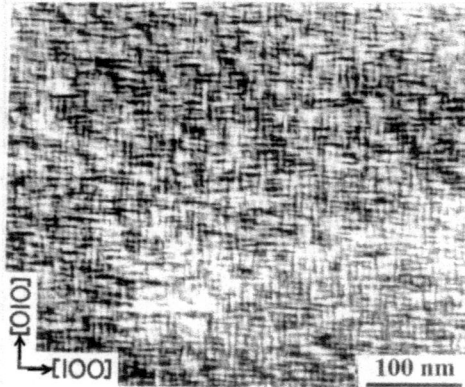

Fig. 2.18 TEM of microstructure of spinodal decomposition in Fe-Mo alloy. (Courtesy by Dr. T. Miyazaki.)

2.2 Heat Treatment in Steels

2.2.1 Fe-C phase diagram

Steels are classified into plain steels and special steels. Plain steels are binary alloys containing only Fe and carbon (C) except impurities unintentionally involved. In special steels alloying element(s) is intentionally added to plain steels to improve the properties. For this reason, special steels are also called alloy steels.

In what follows the phase diagram of Fe-C binary system, the fundamental to understanding properties of steels, will be discussed. Figure 2.19 indicates the phase diagram of Fe-C binary system. Usually Fe-C binary phase diagram contains two phase diagrams, one is genuine equilibrium phase diagram between Fe and graphite (C) and the other is meta-stable phase diagram between Fe and cementite (an intermetallic compound between Fe and C, chemical formula being Fe_3C, denoted by θ).

From an academic point of view, the genuine equilibrium phase diagram (Fe-C) is to be expressed in solid lines and the meta-stable phase diagram (Fe-$Fe_3C(\theta)$) be drawn in broken lines. However, from the practical viewpoint the meta-stable phase diagram is far more important than the genuine equilibrium one, so that Fe-θ diagram is drawn in solid lines and Fe-C(graphite) in broken lines.

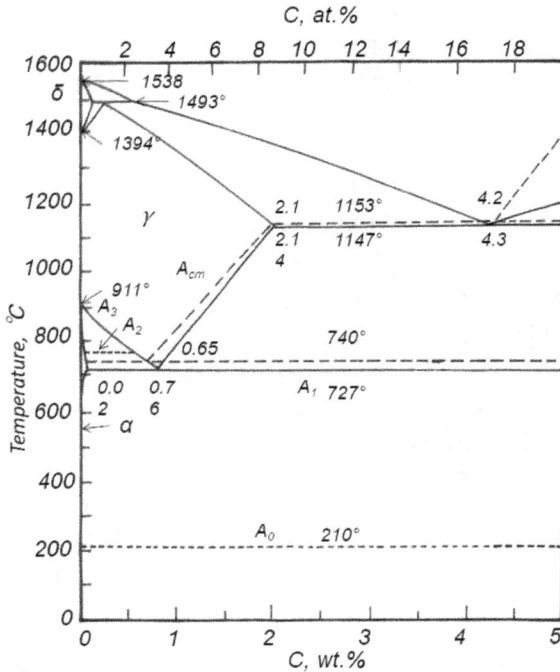

Fig. 2.19 Fe-C phase diagram. Solid lines are for Fe-θ meta-stable equilibrium system and broken lines are for Fe-C equilibrium system.

Terminology used in the phase diagram of Fe-C system.

A_0: Curie point of magnetic transformation of cementite (θ-Fe$_3$C)) (260°C)
A_1: Eutectoid temperature austenite (γ) $\leftrightarrows$ pearlite ($\alpha+\theta$)
    Ac_1: Onset of A_1 transformation on heating, i.e. pearlite ($\alpha+\theta$) $\rightarrow \gamma$
    Ar_1: Onset of A_1 transformation on cooling, i.e. $\gamma \rightarrow$ pearlite ($\alpha+\theta$)
    Ae_1: Equilibrium temperature of A_1 transformation (727°C)
A_2: Curie point of magnetic transformation of ferrite (α-Fe)
A_3: Transformation temperature of $\gamma \leftrightarrows \alpha$ (for pure iron 911°C, decreasing
    with carbon content)
    Ac_3: Onset of A_3 transformation on heating, i.e. $\alpha \rightarrow \gamma$
    Ar_3: Onset of A_3 transformation on cooling, i.e. $\gamma \rightarrow \alpha$
A_4: Transformation temperature of $\gamma \leftrightarrows \delta$ (1390°C)
A_{cm}: Transformation temperature of cementite (θ-Fe$_3$C))$+\alpha \leftrightarrows$ austenite (γ).
    Applied only to hypereutectoid steel.

Here, γ phase is referred to as austenite (named after William Chandler Roberts-Austen[9]) with FCC structure, α and δ phases are referred to as ferrite[10] and identical phase with BCC structure. δ phase occurs at very high temperatures and its reaction is quite rapid. In the practical heat treatment γ (austenite) phase and α (ferrite) phase play far more important roles. Solubility limit of C in γ phase is 2wt.% in a solid solution, while that in ferrite is only 0.025wt%. When γ is cooled slowly, eutectoid reaction, $\gamma(FCC) \rightarrow \alpha\text{-Fe}(BCC) + Fe_3C(\theta)$, takes place at the eutectoid temperature 727°C and a mixture of α-Fe and θ is formed. This structure is referred to as pearlite.[11] Carbon atoms dissolved in the γ phase are expelled from γ phase, and react with Fe atoms to form Fe_3C, which precipitates. In order for this to occur, a long-range motion (diffusion) of C is required. Figure 2.20(a)(b) show SEM and TEM micrographs of pearlite.

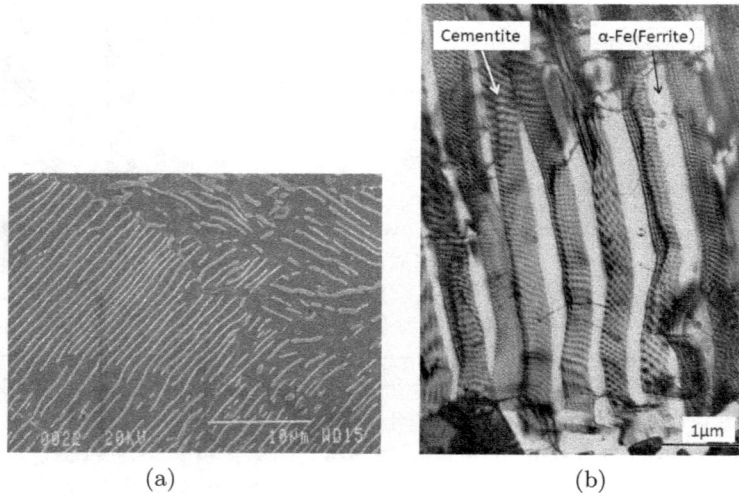

(a)   (b)

Fig. 2.20 (a) SEM of pearlite (courtesy of Nippon Steel). (b) TEM of pearlite. Fringe patterns on cementite are moiré patterns caused by interference of electron beam and have nothing to do with real structures (courtesy of Dr. M. Nemoto).

---

[9]William Chandler Roberts-Austen (1843–1902), UK.

[10]A type of ceramic compound, composed of Fe_2O_3 and additional metallic elements, is also called ferrite.

[11]In those days when pearlite was found, the resolution of an optical microscope was too poor to resolve pearlite into α-Fe and cementite. As a result, *pearlite* appeared as a whole pearl-like tint. This is the reason why the mixture of α-Fe and cementite was named pearlite.

When cooled more slowly, graphite precipitates through eutectoid reaction as the equilibrium phase instead of meta-stable cementite at 740°C. This eutectoid structure between α-Fe and graphite is referred to as ledebrite.

| cast iron |

Iron containing excess of 1.7wt.% C is referred to as cast iron. As is obvious from Fig. 2.19 the melting point decreases with increasing C content: The melting point of pure iron is 1538°C, while at 4.3wt.%C it is reduced to 1500°C. This makes it easier to melt, therefore suitable for making castings.

Cast irons are classified into white cast iron and grey cast iron. White cast iron consists of α-Fe and cementite (Fe_3C), and the fracture is white. Grey cast iron contains ~2wt.%Si, and Fe-cementite system becomes unstable and stable Fe-graphite system occurs. When grey cast irons are cut with a saw, the debris contains graphite. Graphite is a core of pencil and the cross section appears grey. Graphite is soft and a good lubricant, suitable for sliding surfaces of a lath and milling machines.

In general graphite is a flake, so that the existence of graphite flake in α-Fe corresponds to the existence of a crack. Addition of cerium and magnesium changes the morphology of graphite from flake to sphere: The ductility is much improved. Such cast irons are referred to as spheroidal graphite cast irons.

2.2.2   Quenching and martensite

a. Bain's relationship

Let us consider the behavior of C when γ-phase is cooled rapidly (quenched). Quenching γ-phase results in a remarkable hardening. This is referred to as quench hardening and is fundamental to heat treatments of steel.

Mechanisms for quench hardening are as follows. During quenching, before diffusion of C is not completed, phase transformation of γ(FCC)$\rightarrow \alpha$-Fe (BCC) is completed.[12] As a result, C atoms are trapped within α-Fe (BCC) without precipitation as Fe_3C. Such a transformation without accompanying diffusion is referred to as diffusionless transformation or martensitic transformation (named after Martens[13]).

---

[12]Strictly speaking, as will be discussed in Sec. 2.2.2b, γ cannot transform into BCC directly but into BCT (body-centered tetragonal): based on BCC but the length of the c axis is different from that of a axis (in this case longer).

[13]Adolf Martens (1850–1914), Germany.

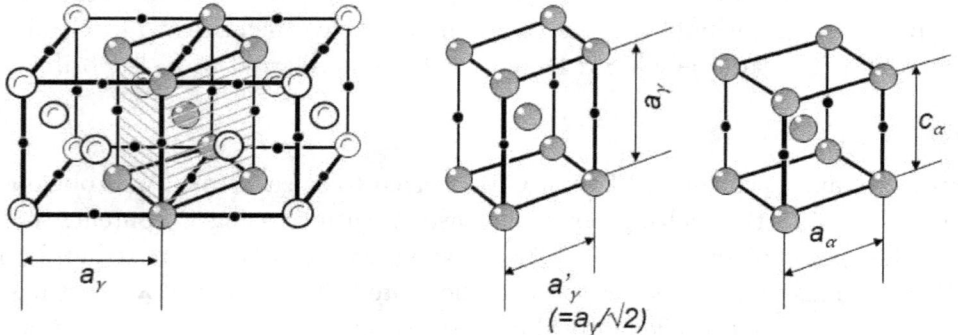

Fig. 2.21 Bain's strain. (a) indicates structure of austenite (FCC). Hatched area in (a) is extracted in (b). Compressing the lattice vertically in (b) and extending horizontally leads to BCC structure shown in (c). Large ○ and ◉ denote Fe atoms, while smaller ● denotes carbon atoms.

Let us consider crystallography in transformation from FCC→BCC in Fe. As can be seen from Fig. 2.21 when two unit-cells of FCC are arranged face-to-face, a BCT lattice hides itself inside. If this BCT lattice is compressed along the c-axis in such a way that the lattice constant c is equal to the lattice constant a, the BCT transforms into BCC. This is referred to as Bain's[14] relationship (or strain).

b. Role of carbon

In the binary Fe-C system at high temperature, C atoms are present as a solid solution in γ-Fe. C atoms are small enough to dissolve in Fe interstitially to form an interstitial solid solution. The position occupied by interstitial C in γ-Fe (referred to as austenite: FCC) are shown by ● in Fig. 2.21(a) (octahedron site; the apexes of the octahedron are occupied by Fe). In austenite (FCC) C is distributed equally at positions indicated by ●, thus the distribution probability along crystallographically equivalent a, b and c axes should be equal. However, once FCC→BCC structural transformation occurs according to Bain's strain, C occupies preferentially at position ● along the c-axis. As a result, the c-axis is elongated more than a and b axes. That is, FCC transforms into BCT (not BCC). The ratio of c-axis to $a(b)$ axis is referred to as tetragonality and is denoted by c/a. With increasing C concentration in γ-Fe solid solution, the amount of C atoms occupying ● along the c-axis after martensitic transformation increases, resulting in an increase in the length of c-axis (Fig. 2.22).

---

[14]Edgar Collins Bain (1891–1971), USA.

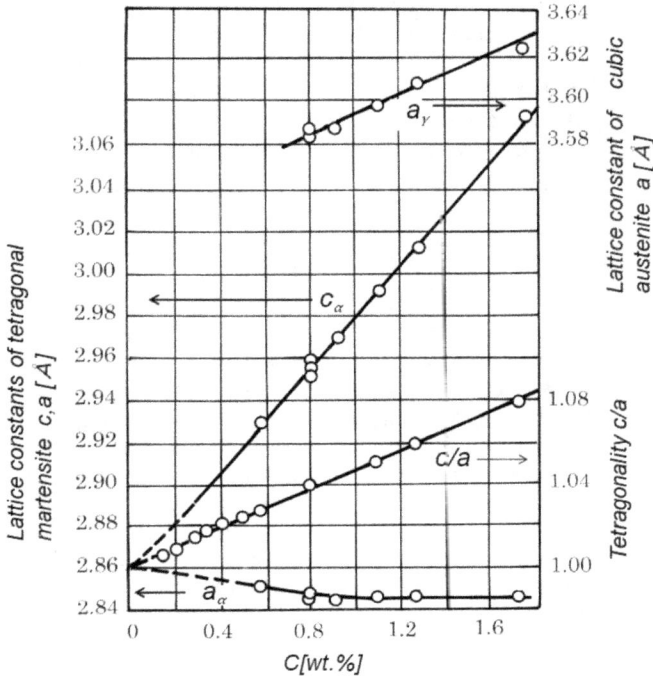

Fig. 2.22 Dependence of lattice constants of martensite and austenite on carbon content (reproduced from *Kinzokuzairyoukisogaku* (in Japanese), edited by Ozaki *et al.* Asakura, 1978, Fig. 7-6 (p. 131)).

This implies that α-Fe, which has BCC structure, is enforced to have BCT since it is supersaturated with C. Thus, with increasing C content, c/a and hence the strain increases, and the hardness increases as well.

Exercise. Confirm that positions of ● in Fig. 2.21(a) are in the centers of octahedra.

c. Features of martensitic transformation

Features of martensitic transformation can be summarized as follows:

(1) It is a reversible transformation from a single phase to another single phase ($\gamma \to \alpha$). By contrast, precipitation is a transformation from a single phase to multiple phases.
(2) It is not accompanied by diffusion. In other words, atoms at the lattice points in the matrix phase cooperatively move to new martensitic phase

(Fig. 2.23(a)→(b)). In other words, the transformation is achieved by shear, without accompanying change in composition.[15]

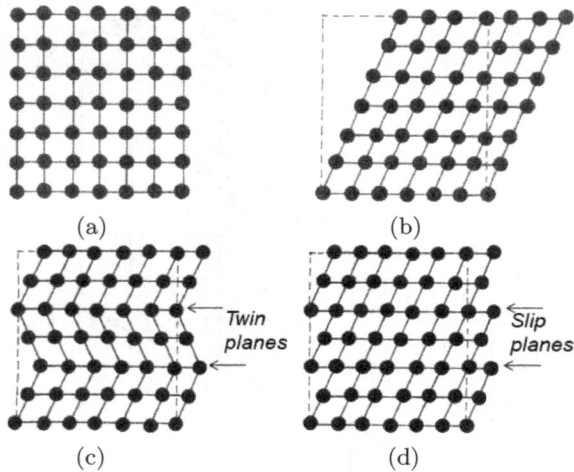

Fig. 2.23 Deformation by slip (b), (d) and by twinning (c). (a) indicates the matrix and (b) martensite. Outer shapes change considerably. In order to accommodate the change in outer shape twinning (c) or slip (d) is necessary.

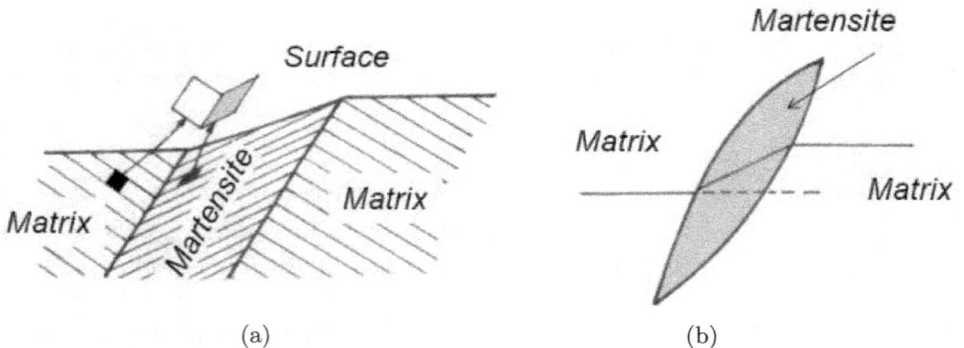

Fig. 2.24 Geometrical relationship between a martensite and the matrix. (a) Cross section. (b) Plan view.

(3) As a result, a surface relief is formed (Fig. 2.24(a)).

(4) A martensite is nucleated on a specific crystallographic plane of the matrix. This plane is referred to as a habit plane.

(5) Between the matrix and a martensite exists a crystallographic orientation relationship (Fig. 2.24(a)).

---

[15]In this respect, Bain's relation does not conform to the features of a martensitic transformation.

Fig. 2.25 Microstructure of a martensite in a steel. Straight contrasts running from left upper to right lower are twins. Finer structures inside twins are dislocations. The density of dislocations is so high that individual dislocations are not resolved.

(6) Martensite contains a high density of lattice defects. As the crystal structure changes on martensitic transformation, so does the overall outer shape of the crystal. Practically in the case of a polycrystal, the newly transformed martensite is restricted with surrounding crystal grains. Martensite is enforced to deform plastically to accommodate itself. Plastic deformation is achieved by slip due to dislocations and/or deformation twin,[16] depending on the composition of the steel under question. Figure 2.25 shows an example of the microstructure of martensite in a steel; it contains a high density of dislocation and deformation twin.

---

[16]Figure 2.23(c) shows a twin plane schematically. Obviously, there is a mirror symmetry between upper and lower crystals facing the **twinning plane**. Figure 2.26(a) shows a high-resolution transmission electron micrograph of a deformation twin in Si. M and M′ are matrices and T is a twin. *12-23-34* is a (doubly bent) lattice plane. Between T and M lattice planes *23* and *34* are situated as if C_1C_2 is a mirror (mirror symmetry). The interface between T and M coincides with C_1C_2. By contrast, between T and M′ while lattice planes *12* and *23* is in mirror symmetry, the interface between M and T is not completely continuous but discontinuous at positions indicated by arrows. That is, the overall interface is not mirror symmetry. An interface like C_1C_2 is referred to as a **coherent twin interface**, and that like I_1I_2 as an **incoherent twin interface**. This difference results from variation of thickness of twin T. Note that the thickness of T decreases from upper to lower.

Besides a deformation twin, an annealing twin can be formed during annealing. Annealing twins occur in FCC metals (Cu, Ag, Ni, austenitic stainless steel). On the other hand, deformation twins occur in BCC(Fe), HCP (Mg, Ti, Cd, Zn) and tetragonal β-Sn. Annealing twins are much coarser than deformation twins (Fig. 2.26(b)).

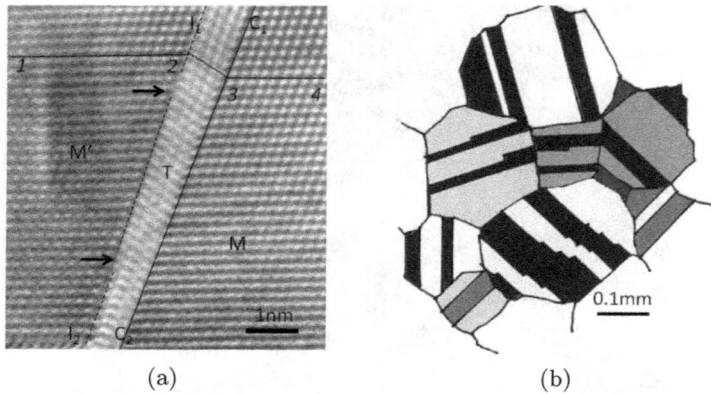

(a)   (b)

Fig. 2.26   HREM of a deformation twin in Si (a). Annealing twins (schematic).

d. Crystallographic relationship

Martensite is closely linked with its matrix γ phase through an orientation relationship. Orientation relationship defines crystallographic planes and/or directions common to both phases. For instance, for Bain's relationship,

$(001)_{fcc}//(001)_{bcc}$

$[110]_{fcc}//[100]_{bcc}$

holds. However, this relationship has not been observed experimentally in steel. The following relationships are confirmed experimentally:

1) **Nishiyama's relationship**

$(111)_{fcc}//(011)_{bcc}$

$[\bar{2}11]_{fcc}//[01\bar{1}]_{bcc}$

2) **Kurdjumov–Sacks relationship**

$(111)_{fcc}//(011)_{bcc}$

$[\bar{1}01]_{fcc}//[\bar{1}1\bar{1}]_{bcc}$

Exercise.   (111) plane of γ-Fe (FCC; $a = 3.5852$ Å) and (011) plane of α-Fe (BCC; $a = 2.8661$ Å) are shown in Fig. 2.27. Based on this, discuss Kurdjumov–Sacks and Nishiyam's relationships.[17]

---

[17]Both in Nishiyama's and Kurdjumov–Sacks relationships $(111)_{fcc}//(011)_{bcc}$ is common. On the other hand, an angle between $[\bar{1}1\bar{1}]$ and $[11\bar{1}]$ in FCC is $70°32'$, while the corresponding angle between $[\bar{1}01]$ and $[101]$ in BCC is $60°$, the gap being as much as $10°32'$. There is a difference between Nishiyama's and Kurdjumov–Sacks relationships in how to accommodate this gap $10°32'$. In Nishiyama's relationship, this gap is shared equally by the two directions $[\bar{1}1\bar{1}]$ and $[11\bar{1}]$. In Kurdjumov–Sacks relationship one out of the two directions $[\bar{1}1\bar{1}]$ and $[11\bar{1}]$ accommodates the gap, with the other free from accommodation totally. Nishiyama's relationship is egalitarianism, while Kurdjumov–Sacks relationship is disparity?

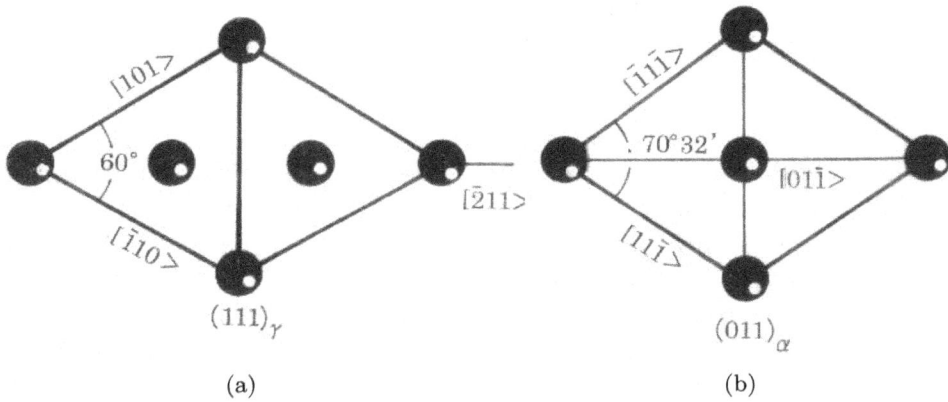

Fig. 2.27 (a) (111) plane of γ-Fe (FCC, $a = 3.5852$ Å). (b) (011) plane of α-Fe (BCC, $a = 2.8661$ Å).

3) Greninger–Troiano relationship

$(111)_{\text{fcc}}$ and $(011)_{\text{bcc}}$ are almost parallel with $1°$ inclined with respect to each other.

$[\bar{1}01]_{\text{fcc}}$ and $[\bar{1}1\bar{1}]_{\text{bcc}}$ are inclined by $2.5°$. This relationship is between Nishiyama's and K-S relationships.

4) Shoji–Nishiyama's relationship

In Co-Ni, Fe-Mn, Fe-Ni and Fe-Ni-Cr alloys, martensite with HCP structure occurs instead of BCT, and is referred to as ε-martensite as compared to α-martensite with BCC or BCT structure. ε-martensite contains fine twins and the orientation relationship is

$[\bar{1}01]_{\text{fcc}}//[11\bar{2}0]_{\text{hcp}} = [110]_{\text{hcp}}$[18]

$(111)_{\text{fcc}}//(0001)_{\text{hcp}} = (001)_{\text{hcp}}$.

This is referred to as Shoji–Nishiyama's relationship.

---

[18]Crystallography of HCP: Indices based on the three axes a_1, a_2 and c (for instance [110] or (110)) are generally used in hexagonal crystal. Four indices like $(11\bar{2}0)$ are often used. For detail of conversion between three and four indices refer to Appendix.

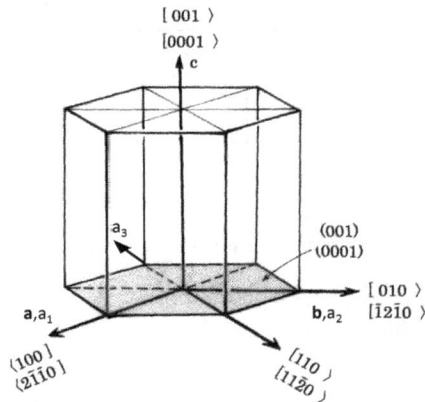

Exercise. Confirm that arrangements of atoms on (111) plane in FCC structure is the same as that of (0001) plane in HCP structure.

e. Habit plane

In general martensite assumes a plate-like shape and occurs along a specific crystallographic plane of the matrix. This plane is referred to as a habit plane and expressed by the crystallographic plane in the matrix. An interface between martensite and the matrix shown in Fig. 2.24(a) is the habit plane. Occasionally martensite is lenticular as shown in Fig. 2.24(b) whose habit plane cannot be identified. Usually there exists a linear feature called a midrib. The habit plane is defined using this midrib.

f. Retaining austenite and deformation-induced martensite

Martensitic transformation from γ accompanies a big strain, so that martensitic transformation on cooling and reverse transformation on heating are accompanied with a hysteresis (Fig. 2.28). Starting and finishing temperatures of martensitic transformation on cooling are defined as M_s and M_f, respectively. Starting and finishing temperatures of the reverse transformation from martensite to austenite on heating are defined as A_s and A_f, respectively.

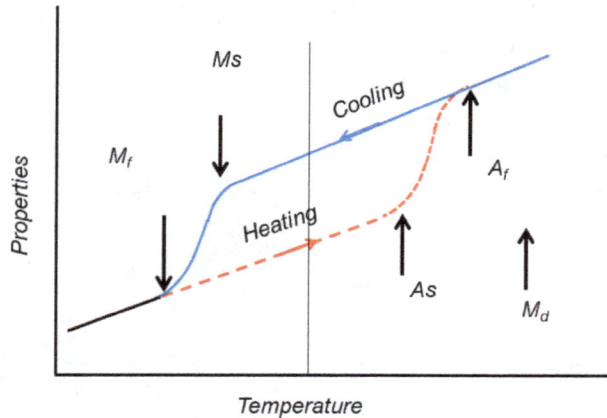

Fig. 2.28 Hysteresis loop in a martensitic transformation. M_s is the starting point of martensitic transformation on cooling and M_f is the finishing point. A_s is the starting point of reverse martensitic transformation from α to γ on heating and A_f is the finishing point. M_d is the upper limit of temperature below which a martensitic transformation can be induced by deformation.

Not all γ grains transform into martensite, since preceding martensite grains prevent the retaining γ from further transformation. Austenite which fails to transform into martensite is referred to as retained austenite. When the retained austenite is abundant, the quenched steel exhibits less hardening. A number of technique to reduce the amount of retained austenite has been invented:

1) *Alloying*; Alloying elements can reduce the amount of retained austenite. This is called improvement of hardenability.

2) *Subzero treatment*; Holding at temperature below room temperature enhances martensitic transformation.

3) *Plastic deformation*; Martensitic transformation is essentially shearing of crystal, so that an external force which enhances this shearing aids the martensitic transformation. This phenomenon is referred to as strain-induced martensitic transformation.[19] Strain-induced martensitic transformation can take place at temperatures in excess of A_f point. However, if the temperature is too high, the strain-induced martensitic transformation does not take place. The critical temperature below which the strain-induced martensitic transformation takes place is defined as M_d point.

2.2.3   Other martensitic transformations

Although martensitic transformation was named originally after Adolf Martens for $\gamma \rightarrow \alpha$ in iron and steel, diffusionless transformation in other alloy systems is referred to as martensitic transformation in general, a typical example of which is shape memory effect or pseudoelasticity.

a. Shape memory effect and pseudoelasticity

Figure 2.29 shows a stress-strain curve of a conventional alloy which does not show pseudoelasticity nor shape memory effect. Figure 2.30 shows a stress-strain curve obtained in tension of an alloy which shows pseudoelasticity and shape memory effect. Here, stress (σ) is a force applied to the specimen divided by the cross-sectional area of the specimen and expressed in kg/mm^2 or MPa (1 kg/mm$^2 \fallingdotseq 10$ MPa). Strain (ε) is an increment in length divided by the original length of the specimen. Suppose that a specimen with the

---

[19]Hadfield steel containing Fe-10%Mn was invented by an English metallurgist, Robert Hadfield. Austenite obtained by quenching from high temperatures is moderately hard. However, any attempt to cut or abrade the surface induces local transformation, so that the steel cannot be machine cut by ordinary cutting tools. For this reason, Hadfield steel is used as a shovel of a shovel car.

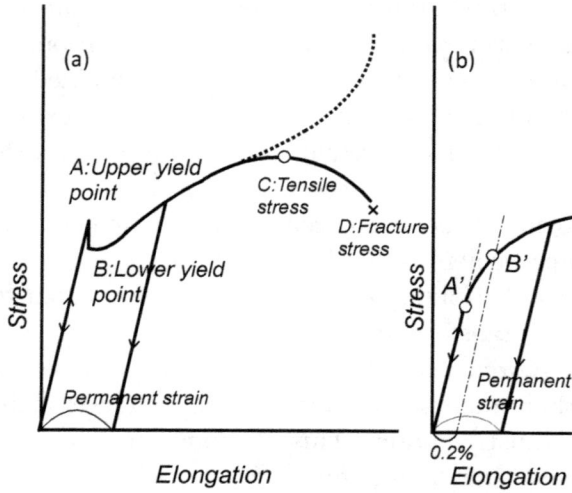

Fig. 2.29 Stress-strain curves of normal metallic materials. (a) is for steel and (b) is for non-ferrous materials. Dashed line in (a) indicates a true stress.

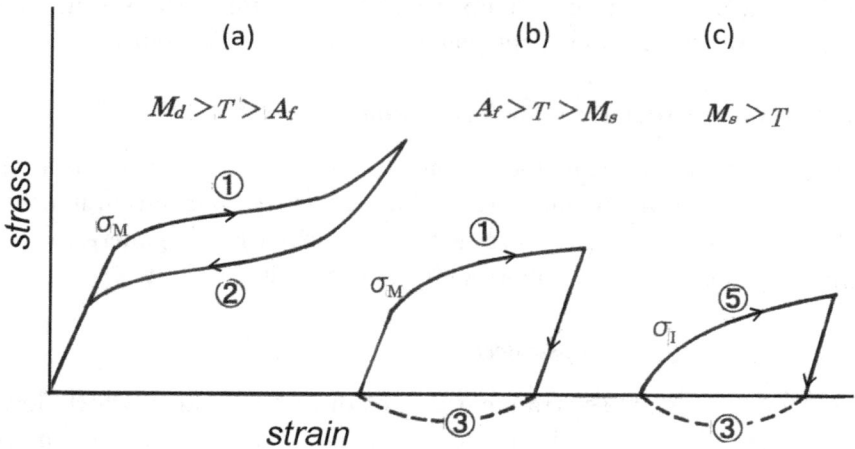

Fig. 2.30 Stress-strain curves of shape memory alloys. (a) indicates pseudoelasticity. (b) and (c) indicate shape memory effects. For ①, ②, ③ and ⑤ refer to Fig. 2.31.

original length of 10 mm is elongated to 11 mm, then the strain ε is given by $\varepsilon = (11 - 10)$ mm/10 mm $= 0.1(10\%)$ in the non-dimensional case.

In Fig. 2.29, stress is proportional to strain when the stress (strain) is small. This is referred to as Hooke's law, and the phenomenon is referred to as elastic deformation. This law holds for all materials when stress is small. On further increase in stress, the linear relationship between σ and ε breaks down, and the strain increases perpetually. This phenomenon is

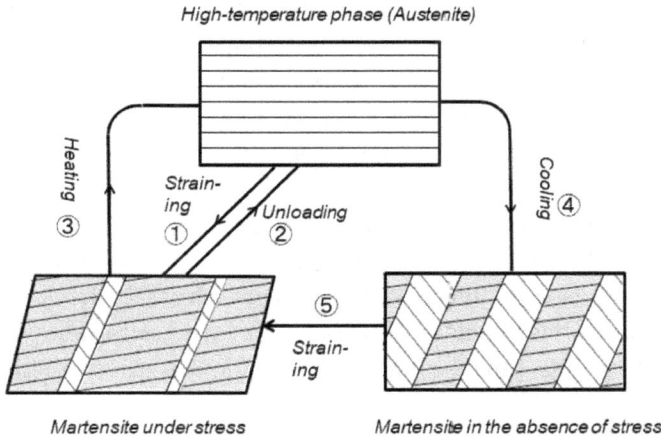

High-temperature phase (Austenite)

Heating

Cooling

Strain-
ing
①

Unloading
②

③

④

⑤

Strain-
ing

Martensite under stress

Martensite in the absence of stress

Fig. 2.31 Mechanisms of shape-memory effect and pseudoelasticity. For ①, ②, ③ and ⑤ refer to Fig. 2.30.

referred to as yield. Figure 2.29(a) shows a stress-strain curve of steel: the stress decreases sharply, followed by a gradual increase in stress. This is referred to as yielding phenomenon. By contrast a stress-strain curve of non-ferrous material such as Cu deviates gradually from the linearity as shown in Fig. 2.29(b). In this case, the yield stress is not defined uniquely. A stress when deviation from the Hooke's law starts (A′) is defined as an elastic limit or a proportional limit. In engineering, a stress corresponding to a strain of 0.2%(B′) is defined as a proof stress, which can be used as a measure of yield stress. In either case, a strain increases rapidly after yield and a permanent strain remains after the force is completely unloaded. This is referred to as plastic deformation.

Now in Fig. 2.30, linearity between σ-ε breaks down, yet on unloading a stress-strain curve retraces that on loading without leaving a permanent strain. This phenomenon is referred to as pseudoelasticity or super elasticity. The reason why a prefix "pseudo" is attached is as follows:

1) On complete unloading a permanent strain does not remain.

2) But, proportionality between σ and ε does not hold.

This seemingly magical phenomenon is accounted for by martensitic transformation. Let us discuss this phenomenon based on Fig. 2.31. On loading at temperature $A_f < T < M_d$ a strain-induced martensitic transformation takes place. As a result, the outer shape of the alloy is changed (corresponding to path ① in Fig. 2.31). Here σ_M is a stress required to induce the martensitic transformation. However, under this state martensite is stable only under stress ($\because T > A_f$). Therefore, upon unloading the martensite transforms reversely back to the original high-temperature phase (corresponding to austenite) and the outer shape returns to the initial shape (corresponding to ② in Fig. 2.31). This is referred to as pseudoelasticity.

When $M_s < T < A_{f'}$ the process on loading is the same (path ①). However, in this case the martensite is stable in the absence of strain ($\because T < A_f$). Thus, on unloading the martensite phase remains without reverse-transforming back to the high-temperature phase; the change in outer shape is held. However, on heating this martensite to above A_f the martensite becomes unstable and transforms reversely back to the initial high-temperature phase (path ③ in Fig. 2.31), the outer shape of the alloy returning back to the initial shape prior to deformation. This is a shape memory effect.

When $T < M_s$, martensite is already formed via path ④. Martensite phase in a shape memory alloy contains a lot of twins inside. The martensite formed by quenching contains a pair of variant at almost equal frequency. However, on loading a variant which responds to external stress invades the counterpart (path ⑤). Here, σ_I is a stress required to move an interface between the variants and much smaller than σ_M. On complete unloading the outer shape does not return to the initial shape. However, on heating to $T > A_f$ the martensite reverse-transforms to the high temperature phase, so does the outer shape (path ③ in Fig. 2.31).

b. Toughening of ceramic

Figure 2.32 shows the phase diagram of ZrO_2-Y_2O_3 system. Here, a tetragonal phase (T) corresponds to austenite, a monoclinic phase (M) to α-BCC phase.[20] C stands for a cubic phase. Y_2O_3 containing in excess of 8mol% is stable in the whole range of temperature, so that $T \to M$ transformation does not take place. Such a zirconia alloy is referred to as a stabilized zirconia. Zirconia alloy containing Y_2O_3 in a range of 1.6%∼8%mol consists

---

[20]For definition of C (cubic), T (tetragonal) and M (monoclinic) structures, refer to figures below. It may be easy to remember if you consider that M for monoclinic corresponds to M for martensite. In general, the crystal structure has a higher symmetry at higher temperature. (a) C, (b) T and (c) M.

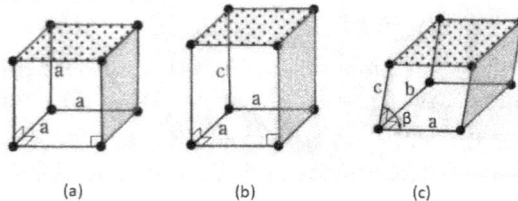

(a)   (b)   (c)

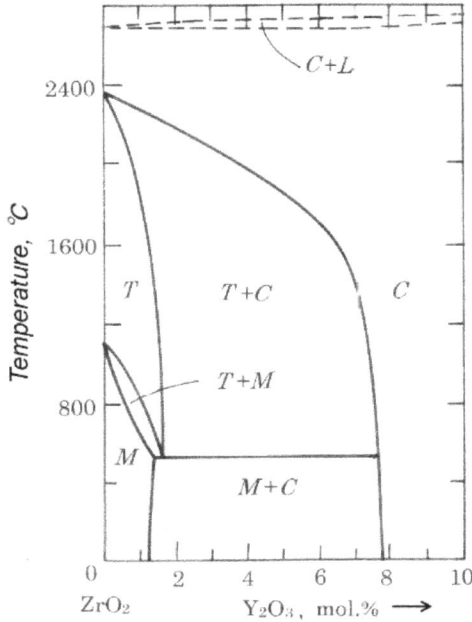

Fig. 2.32 Phase diagram of ZrO$_2$-Y$_2$O$_3$ system (reproduced with permission from S. P. Ray, R. C. Hink and V. S. Stubican, *J. Amer. Ceramic Soc.*, 61(2006) 17).[21]

of two phases, i.e, $T + C$. This alloy contains C phase, thus is stabilized to some extent, and referred to as a partially stabilized zirconia. When T phase is cooled rapidly into M-phase region, diffusionless martensitic $T \to M$ transformation takes place and the zirconia alloy is toughened. However, when a complete M phase is quenched, strain associated with martensitic transformation and/or thermal stress induces cracking. Practically the amount of T phase which transforms into M phase is adjusted using a partially stabilized zirconia.

2.2.4 Tempering of martensite, Temperature-time-transformation (TTT) diagram, Continuous cooling transformation (CCT) diagram

a. Tempering of martensite

In martensitic transformation C atoms are trapped inside a supersaturated solid solution. C atoms try to exit from the supersaturated solid solution.

---

[21]Unit of the horizontal axis is mol.% and indicates the ratio of amount of molecules of ZrO$_2$ and Y$_2$O$_3$ (which corresponds to at.% for metallic materials).

In order for this to occur, C atoms must diffuse. The diffusion is enhanced at high temperatures. Thus, when a quenched steel is heated a little bit (tempered), C atoms start exiting from the supersaturated martensite and eventually (metastable) cementite (Fe_3C, θ) is formed and the tetragonal martensite changes into cubic α-Fe(ferrite). Concurrently with this, hardness decreases and ductility and fracture toughness increase.

Processes of tempering of martensite

Tetragonal martensite		$\rightarrow$	α-Fe(ferrite)	+	cementite(Fe_3C,θ)
Hardness	High	$\rightarrow$	Low		
Ductility	Low	$\rightarrow$	High		

In practice, at the intermediate stages, cubic martensite and $Fe_2C(\varepsilon)$ phases are formed. Tempering proceeds in three stages:

Stage I (70~150°C): Very fine carbides precipitate from a supersaturated tetragonal martensite solid solution. This carbide is hexagonal ε-Fe_2C. Matrix martensite changes into cubic martensite, which contains ~0.25wt.%C. This martensite is referred to as low-carbon martensite.

Stage II (250~300°C): Retained austenite decomposes into low-carbon martensite and ε-Fe_2C.

Stage III (270~400°C): ε-Fe_2C changes into θ-Fe_3C and concurrently low-carbon martensite decomposes into α-Fe(ferrite) + θ-Fe_3C. In other words, meta-stable equilibrium is attained. These processes are summarized in Fig. 2.33.

At what stage is a steel used for practical service is the key point of heat treatments, which is determined by purpose, economic reasons and so on.

$\boxed{\textbf{Note}}$ Terminology of heat treatments

quenching
    Rapid cooling. Water quench; WQ, oil quenching; OQ, ice water quench; IWQ and brine quenching.

quench hardening
    When an austenite in a steel is quenched from above Ac_3 point, the steel is hardened due to martensitic transformation.

Quench

High temperature	As-quenched	Stage 1	Stage 2	Stage 3
Austenite (γ-Fe)	Martensite → Low-carbon martensite + ε-Fe₂C	Retained austenite		α-Fe + θ-Fe₃C (Cementite)
Properties	Hard, brittle	←——————————→		Soft, tough

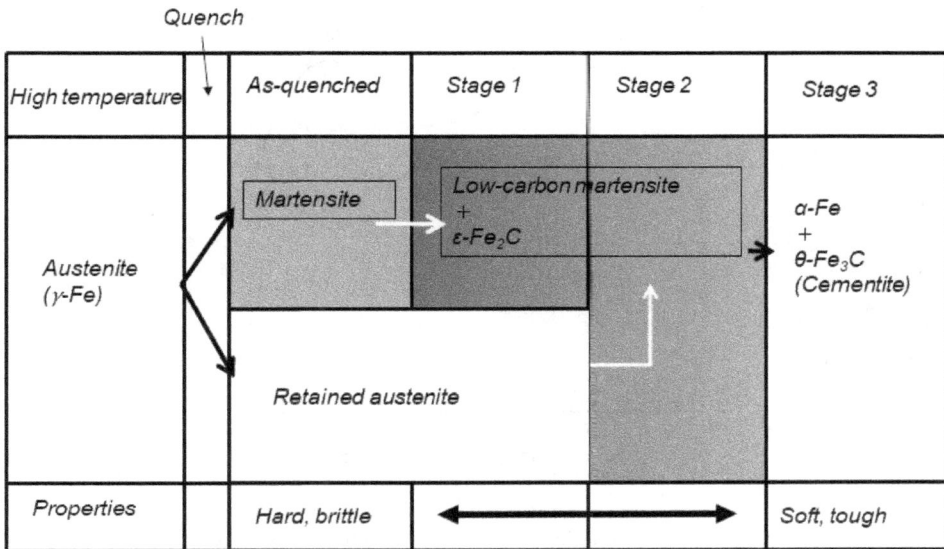

Fig. 2.33   Tempering of a martensite.

normalizing
    A steel is heated at temperatures about 30∼50°C higher than Ac_3 or Ac_m, followed by air cooling (AC).

tempering
    Heating a quench-hardened steel below A_1 in order to decrease hardness and increase fracture toughness.

annealing
    In order to modify microstructure and get rid of internal stress, heating at appropriate temperature and then slowly cooled.

b. Secondary hardening

When martensite in an alloyed steel is tempered, an alloying element and C react to form a carbide which differs both from θ-Fe₃C and ε-Fe₂C, and the hardness increases. Figure 2.34 shows how hardness varies when a Mo-containing steel is tempered depending on the Mo content. In a Mo-free steel the hardness decreases monotonously with tempering temperature, however, in a 0.5%Mo steel, decrease in hardness above 400°C diminishes and in steels containing more than 3%Mo hardness increases with increasing temper temperature, reaching a peak hardness at around 600°C. This phenomenon is referred to as secondary hardening.

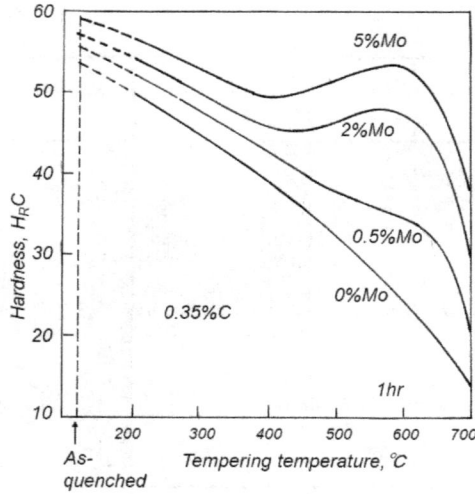

Fig. 2.34   Secondary hardening of Mo-bearing steels.

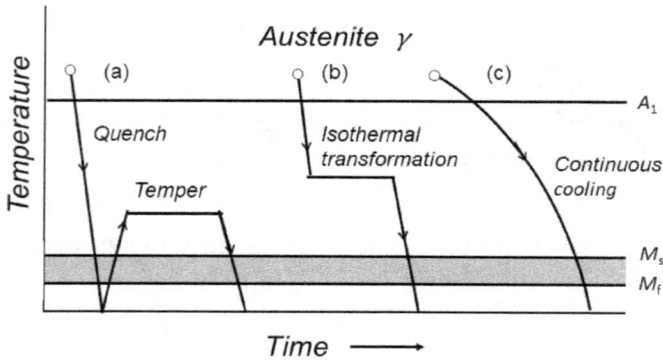

Fig. 2.35   Practical heat treatments of steels.

c. Temperature-Time-Transformation (TTT) diagram

The practical heat treatments of steel are classified into three categories as shown in Fig. 2.36. Tempering (Fig. 2.36(a)) is already described in Sec. 2.2.4.

When manufacturing steels in a steel mill, energy efficiency of quench-temper treatment described in Sec. 2.2.4a is poor, since a lump of steel which is cooled down to room temperature must be heated again for tempering.

If a steel lump is held at a temperature corresponding to the tempering temperature during cooling from high temperatures and its microstructure is controlled, the energy efficiency is improved. Such a heat treatment is

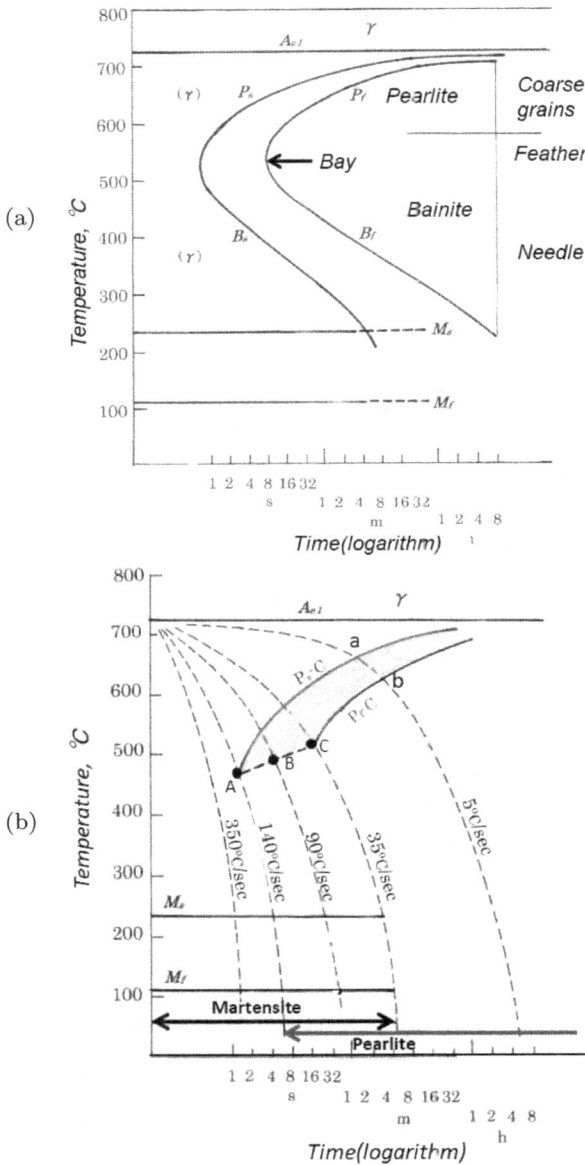

Fig. 2.36 TTT diagram (a) and CCT diagram (b). In (a), P_s and P_f stand for the starting and finishing points of pearlite transformation, respectively. B_s and B_f stand for the starting and finishing points of bainite transformation, respectively. In (b) P_s-C and P_f-C stand for the starting and finishing points of pearlite transformation, respectively. Here C indicates C for continuous cooling. M_s and M_f are the starting and finishing points of martensitic transformation, respectively. When cooled at a cooling rate of 350°C/sec pearlite transformation does not take place, reaching M_s point, where martensitic transformation starts and finishes at M_f point. Thus, the microstructure obtained at room temperature consists of martensite alone. When cooled at less than 35°C/sec, martensitic transformation does not take place and the resultant microstructure consists of pearlite alone. When passing A-C, a mixture of martensite and pearlite is obtained.

referred to as isothermal treatment (Fig. 2.35(b)). TTT diagram is useful to understand what is happening in the isothermal treatment (Fig. 2.36(a)).

Since the steel is held above M_s, what happens here is not a martensitic transformation but a eutectoid (pearlite) transformation fundamentally. That is, α and θ precipitate from a supercooled austenite. The higher the degree of supercooling, the stronger the tendency for precipitation. This desire for precipitation is called the driving force. The degree of supercooling increases with decreasing temperature; in other words, the driving force increases. However, on the other hand, in order for θ to precipitate, C atoms dissolved in a supersaturated solid solution must diffuse to coagulate to some extent. Diffusivity diminishes with decreasing temperature. As a result, at a temperature range where both the driving force and the diffusivity are moderately high, the rate of precipitation of θ is the highest. In a TTT, at such a temperature range a bay appears as shown in Fig. 2.36(a) where the rate is highest.

Above the bay a distinct pearlite is formed, while a microstructure formed below the bay is finer and is called a bainite.

Break

Measurement of hardness is a simple way to estimate mechanical properties. An indenter whose hardness is higher than the sample under question is pressed with a constant force L onto the surface, then the area (L) of the impression of the indent is measured. Hardness is defined by L/S. A variety of hardness is defined depending on geometry and material used as an indenter.

Rockwell (after Hugh M. Rockwell and Stanley P. Rockwell): Indenter is cone of steel (B-scale) ($H_R B$) or spherical diamond (A, C scale) ($H_R A$ or $H_R C$).

Brinnel (after J. A. Brinnel): Indenter is spherical steel (Bn).

Vickers (after a company in UK): Indenter is a pyramidal diamond (Hv or VH (Vickers Hardness), DPH (Diamond Pyramid Hardness))

Knoop (after Frederick Knoop): The geometry of this indenter is an extended pyramidal diamond with the length to width ratio being 7:1. Used to measure the anisotropy of hardness.

Berkovich (after E.S. Berkovich): Indenter is a three-sided pyramid.

Recently, a method called nanoindentation is devised, where a diamond indenter is driven into the specimen surface and the applied force and displacement data are collected dynamically. Material properties are derived from the load and depth data. Specimens are typically relatively smooth and flat.

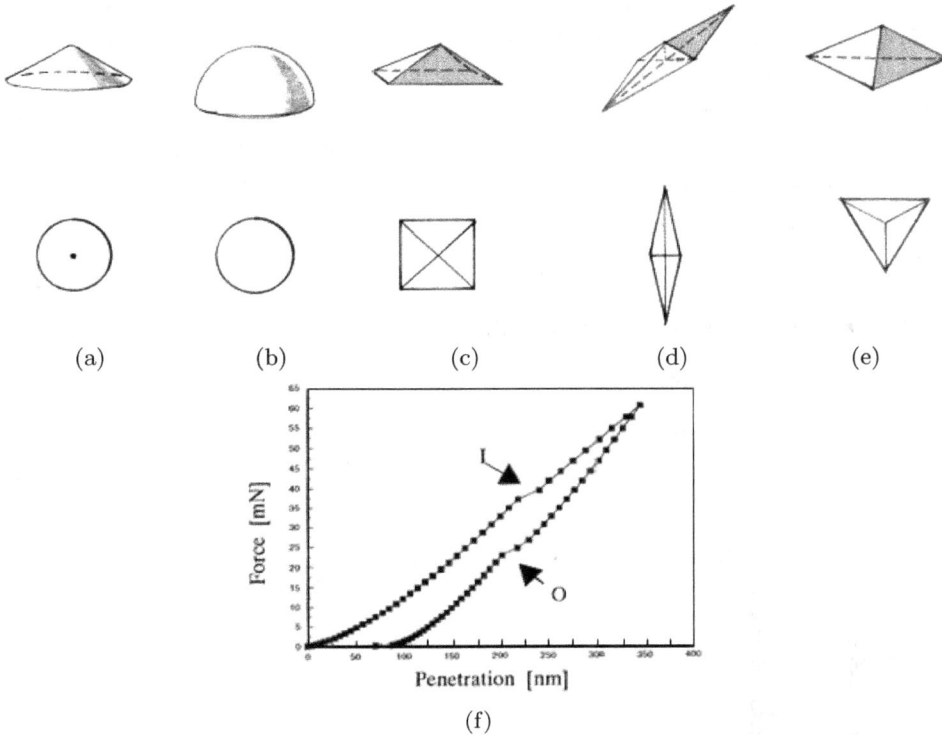

Fig. 2.37 (a–e) Geometry of indenters (upper) and corresponding images of indenters (lower). (f) Nanoindentation using a Brinell type indenter (Si at room temperature) (a) Rockwell (B scale)(H_RB), (b) Rockwell (A and C scales)(H_RA, H_RA) and Brinnel. (c) Vickers, (d) Knoop, (c) Berkovich. (f) Discrete points on the stress-strain curves (I) and (O) are referred to as pop-in and pop-out, respectively.

d. Continuously cooling transformation (CCT) diagram

When a steel is welded, the welded part and it surroundings are cooled continuously after welding. CCT diagram is useful to understand what happens during a continuous cooling (Fig. 2.36(b)).

Here, broken lines show cooling curve. Ps-C (starting point of pearlite reaction @CCT diagram) and Pf-C (f stands for finish) correspond to start and finish of pearlite reaction.

When a steel is cooled at 350°C/sec, the steel is cooled down to room temperature without crossing the Ps-C curve, resulting in formation of pearlite totally. When crossing A~C, it passes Ps-C but not Pf-C. In other words, the pearlite reaction is unfinished, resulting in formation of a mixture of pearlite and martensite.

By contrast, when a steel passes points a and b (at a cooling rate of 5°C/sec), austenite (γ) is supercooled until point a, where pearlite starts to form and finishes at point b. Therefore, microstructure obtained after cooling to room temperature is pearlite only and no martensite is formed. That is, in order to obtain martensite at room temperature, the cooling rate must be faster than 35°C/sec to pass point c at the slowest. The slowest cooling rate required to obtain martensite is referred to as a critical cooling rate.

2.2.5 Improvement of hardenability of steel

a. Jominy end quench test

The term hardenability is used to indicate the depth to which a fully martensitic structure can be obtained. As already described in Sec. 2.2.2f, in a plain

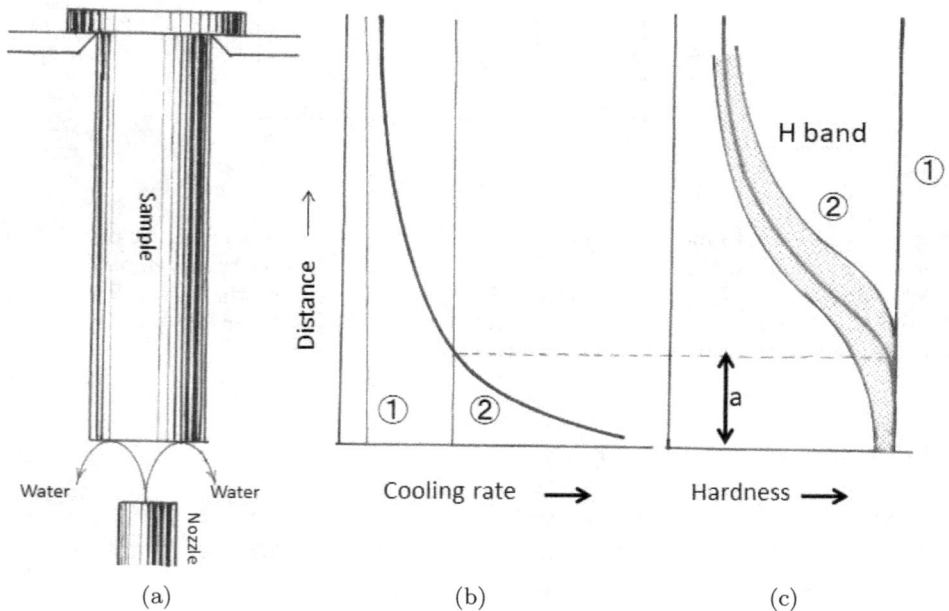

(a)   (b)   (c)

Fig. 2.38 Jominy end quench test. A steel bar heated at high temperature is hanged on a hook, which is cooled by a jet of water (a). The cooling rate is highest at the lower end (b). When a critical cooling rate is low as indicated by ①, the overall sample undergoes a martensitic transformation and hardened (c). When a critical cooling rate is higher (corresponding to ②), a martensitic transformation takes place only in a volume within a distance **a** from the end. The resultant profile of hardness is schematically indicated by ② in (c). Error involved in this method is considerably large, so that hardenability is judged based on whether or not a profile of hardness falls in a hatched area ② in (c). This area is referred to as H band.

steel (Fe-C alloy) not all the γ phases transform into martensite but is partly retained (referred to as retained austenite), resulting in a poor hardenability. A conventional method to estimate hardenability is Jominy end quench test. In this method a hot cylindrical bar is quenched from one end by a water jet, and its hardness is measured as a function of the distance from the quenched end (Fig. 2.38(a)). The cooling rate is highest at the quenched end, decreasing with distance upward (Fig. 2.38(b)). When the critical cooling rate is low enough as shown by ① in Fig. 2.38(b) martensitic transformation occurs from the quenched end to the upper part, the hardness distribution being flat over the full length of sample as shown by ① in Fig. 2.38(c)). In other words, hardenability is high.

By contrast when the critical cooling rate is high (② in Fig. 2.38(b)), martensitic transformation occurs only near the quenched end, say point **a** in (c), so that hardness drops over point **a**: The hardenability is poor. Practically when distribution of hardness lies in a specified region (hatched in (c)), the hardenability is judged as up to standard. The hatched region is referred to as H band.

b. Improvement of hardenability

Addition of a small amount of alloying element is effective to improve hardenability. Cr, Mo and Mn are especially effective. Such an alloyed steel is referred to as a special steel or an alloy steel.

2.2.6   Case hardening

The greatest advantage of steel is that its strength, hardness and ductility etc. can be controlled by heat treatments. On top of this, it is possible to make the sub-surface harder, while keeping the inner core ductile (case hardening). A typical example of such a case hardening is a Japanese sword, which will be described later. Case hardening in modern metallurgy involves flame hardening, induction hardening, laser hardening, carburization, nitriding, carbonitriding and shot peening, etc.

In flame hardening, the surface of a steel is heated by a flame in such a way that a sub-surface reaches austenite region but the inner core is not hot enough, followed by quenching. This results in martensitic transformation only in the sub-surface region. In induction hardening only a sub-surface is heated by induction heating, followed by quenching. In laser hardening a laser beam is used to heat locally, followed by quenching.

In carburization carbon atoms diffuses into the interior of a steel, followed by quenching. As is obvious from Fig. 2.21 hardness of a martensite in the carburized layer increases. The effect of nitrogen on steel is similar to that of carbon. Carbonitriding is a combination of carburization and nitriding.

Shot peening is different from those described above in that a sub-surface is work-hardened[22] by peening with hard particles.

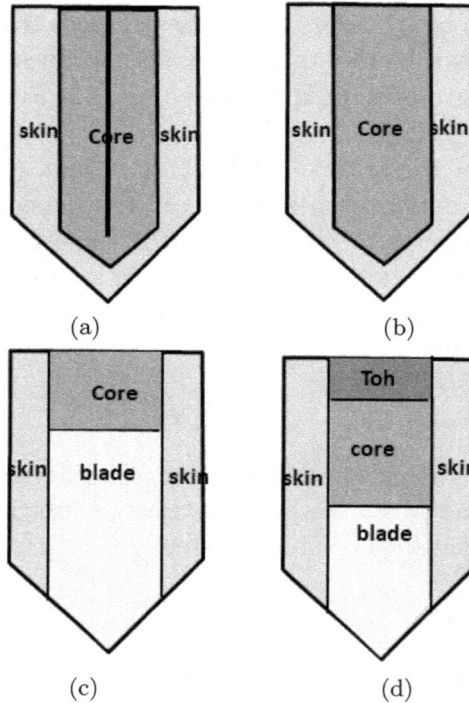

(a)   (b)

(c)   (d)

Fig. 2.39 Cross sections of Japanese sword. (a) Makuritataki, (b) Kobuse, (c) Honsan-mai, (d) Shihozume. (Courtesy of Heibon-sha.)

Japanese swords

A Japanese sword is composed of an inner core (shin) of α-ferrite ($\sim$0.3%C), which is soft and tough, and outer jacket (skin or blade) of martensite ($\sim$0.6%C) (Fig. 2.39). A Japanese sword is essentially different from a

---

[22]When a metal is deformed plastically, it does not recover to the original shape on unloading, leaving a permanent strain (this is the definition of plastic deformation) (see Fig. 2.29). On reloading the stress-strain curve traces the broken line, and the elastic region extends. In other words, a metal hardens. It is an everyday-life experience to bend a metal wire repeatedly to fracture.

conventional case-hardened steel in that the outer jacket and inner core are prepared separately and then laminated together by hot forging. This makes it possible to control the hardness of blade and core independently when quenched.

Before quenching from high temperature a sword is covered with clay in such a way that it is thick on the core and thin on the blade. The thickly painted region (core) is cooled slowly, while the thinly painted region (jacket) is cooled rapidly. As is obvious from Fig. 2.39(b), the thinly painted part (i.e. jacket) transforms into a martensite, while a thickly painted part (core) does not transform into martensite, resulting in soft and tough microstructure. Such a composite material consisting of parts of different properties to achieve better properties as a whole is referred to as a functionally graded material.

Chapter 3

Thermodynamics of Binary Phase Diagrams

3.1 Equilibrium Between Different Phases

The phase diagram is a visualization of equilibrium states among different phases. Therefore, in order to understand phase diagrams properly, it is necessary to have a knowledge of thermodynamics. In Chapter 3 elements of thermodynamics required to understand the binary phase diagram will be given. At equilibrium the free energy of an alloy (system) is minimum. At the same time the chemical potentials of components comprising the binary alloy (system), i.e. atoms of metals A and B (hereinafter denoted by A(○) atom and B(●) atom, respectively) are equal. In what follows it will be demonstrated that these two propositions are equivalent.

3.1.1 Chemical potential

a. Definition of chemical potential

Let us consider an isomorphous (complete) solid solution as shown in Fig. 3.1(a). In Fig. 3.1(b) the Gibbs free energy G of α-solid solution at T_{α}, which is lower than both T_A and T_B (the melting points of metals A(○) and B(●), respectively) is plotted against the concentration of B(●) metal, $x_{\bullet}$. The free-energy/composition curve (G-$x_{\bullet}$ curve) opens upwards.[1]

   The chemical potential of A(○) atom in α-phase with composition $c(= x_{\bullet})$ (expressed by $(\mu_{\circ})_c$) is the intersection of the tangential line at $x_{\circ} = 1(x_{\bullet} = 0)$. Similarly, the chemical potential of B(●) atom in α-phase at composition $c(= x_{\bullet})$ (expressed by $(\mu_{\bullet})_c$) is the intersection of the tangential line of G-$x_{\bullet}$

---

[1]All of the G-$x_{\bullet}$ curves in the rest of this chapter open upwards. Needless to say, the reverse case (that is, opening downwards) is possible and to be discussed later. However, since we are now discussing states with a minimum energy, let us assume temporarily that all the G-$x_{\bullet}$ curves open upwards. The meanings of subscript and superscript are as follows: $(\mu_{atoms}^{phase})_{composition}$.

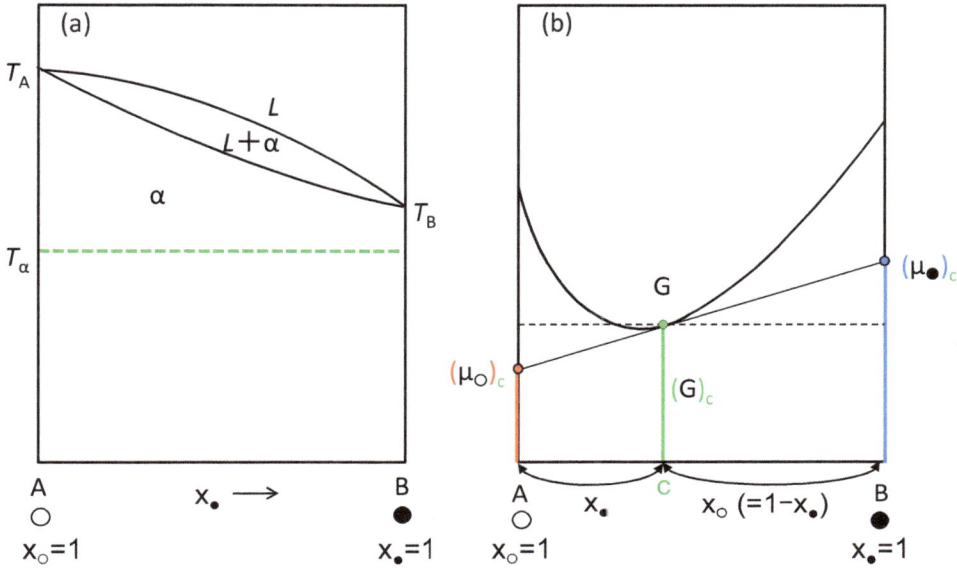

Fig. 3.1 (a) Phase diagram of an isomorphous phase diagram. (b) Relation between Gibbs' free energy (G) and chemical potential (μ) at temperature T_α.

curve at $x_\bullet = 1$). It is obvious that

$$(G)_c = (\mu_\circ)_c \cdot x_\circ + (\mu_\bullet)_c \cdot x_\bullet \tag{3.1}$$

b. Law of common tangential line

Let us consider the case where two phases, α and β, are coexistent in the A($\circ$)-B($\bullet$) binary system (Fig. 3.2). Free-energy/composition curves of α and β phases show minima at composition m and n, respectively. However, α and β phases having the compositions m and n, respectively, are not at equilibrium. The reason for this is as follows: the chemical potential of A($\circ$) atom in α (with composition m) $(\mu^\alpha{}_\circ)_m$ is higher than the chemical composition of A($\circ$) atom in β (with composition n) $(\mu^\beta{}_\circ)_n$. That is, $(\mu^\alpha{}_\circ)_m < (\mu^\beta{}_\circ)_n$. Similarly, for the chemical potentials of B($\bullet$) atom, $(\mu^\alpha{}_\bullet)_m < (\mu^\beta{}_\bullet)_n$.

   At equilibrium between α phase ($x_\bullet = p$) and β phase ($x_\bullet = q$) chemical potentials of A($\circ$) and B($\bullet$) atoms are equal for both in α and β phases.

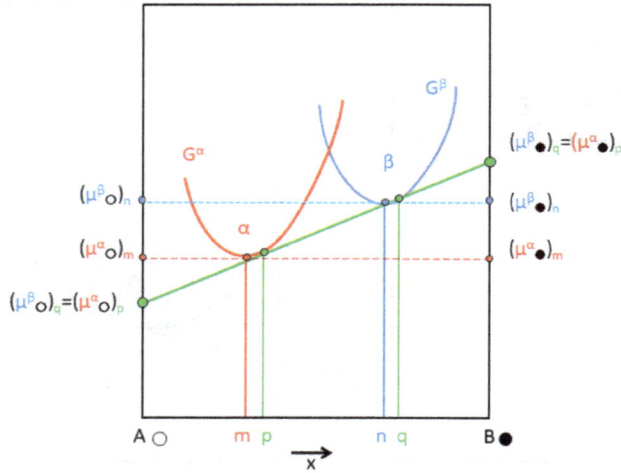

Fig. 3.2 Chemical potentials of α and β phases.

That is,

$$(\mu^{\alpha}{}_{\circ})_p = \left(\mu^{\beta}{}_{\circ}\right)_q \quad \text{for A(○) atom} \tag{3.2a}$$

$$(\mu^{\alpha}{}_{\bullet})_p = \left(\mu^{\beta}{}_{\bullet}\right)_q \quad \text{for B(●) atom.} \tag{3.2b}$$

The compositions of p and q are the points of contact on a common tangential line for free-energy/composition curves of α- and β-phases.

Next, let us consider the case where α-phase with a composition $x_{\bullet} = p'$ and β-phase with a composition of $x_{\bullet} = q'$ are coexistent, where p' and q' are out of equilibrium.

In this case, the chemical potential for A(○) atom is,

$$(\mu^{\alpha}{}_{\circ})_{p'} < \left(\mu^{\beta}{}_{\circ}\right)_{q'}$$

As a result, A(○) atoms migrate from β-phase (with higher chemical potential) to α-phase (with lower potential) just like water flows from a higher position (potential) to a lower position (potential). On the other hand, for B(●) atom

$$\left(\mu^{\beta}{}_{\bullet}\right)_{q'} < (\mu^{\alpha}{}_{\bullet})_{p'}$$

so that B(●) atoms migrate from the higher chemical potential α-phase to the lower chemical potential β-phase. In other words, β-phase is enriched in B(●) atom and α-phase is enriched in A (○) atom.

$$q' \to q$$
$$p' \to p$$

Eventually, the equilibrium compositions (p, q) are reached.

Exercise. Discuss the case shown in Fig. 3.3(b).

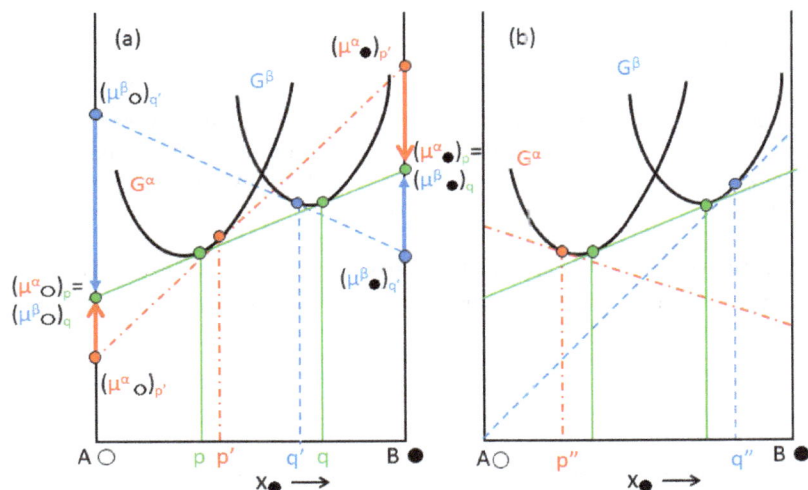

Fig. 3.3 Chemical potentials in α and β phases whose concentrations deviate from the equilibrium ones (indicated in green solid lines).

Now let us discuss from the viewpoint of the total free energy of the alloy system. The free energy of an alloy consisting of a mixture of two different phases $(\alpha + \beta)$ corresponds to position c which sits on a line connecting the free energies of the respective phases $(G_\alpha)_p$ and $(G_\beta)_q$. As is obvious from Fig. 3.4(a) the free energy of a mixture of $(\alpha)_p$ and $(\beta)_q$ at composition c is given by $(G)_c$

$$(G)_c = (G_\alpha)_p \times (\text{amount of } \alpha\text{-phase}) + (G_\beta)_q \times (\text{amount of } \beta\text{-phase})$$

$$= (G_\alpha)_p \times \frac{m}{m+n} + (G_\beta)_q \times \frac{n}{m+n}$$

where p and q are points of contacts on the common tangential line for the free energy-composition lines of α- and β-phases.

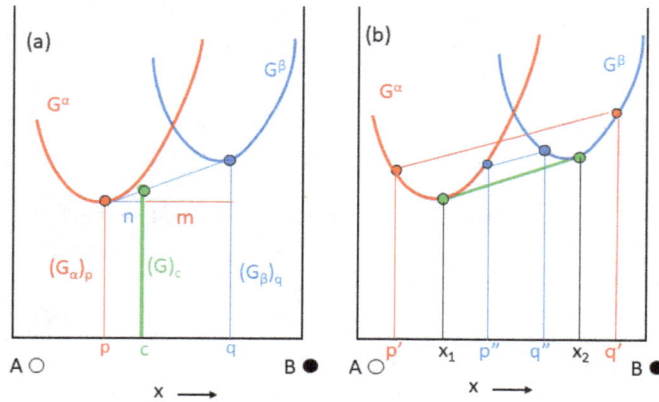

Fig. 3.4   (a) Determination of equilibrium concentrations based on common tangential line. (b) When α and β phases, whose compositions are not at equilibrium, are coexistent, the free energy of a mixture of α and β phases is higher than that corresponding to the equilibrium.

As can be seen from Fig. 3.4(b), any combination of α- and β-phases other than (p, q) such as (p', q') or (p'', q'') results in free energy higher than $(G)_c$.

The foregoing discussions based on chemical potential and free energy reaches the same conclusion, that is,

---

When two phases $(\alpha+\beta)$ are coexistent at equilibrium, their composi-tions, p (of α-phase) and q (of β-phase), correspond to points of contact on the common tangential line of the free-energy/composition curves of α- and β-phases.

When the free-energy/composition curve of α-phase is higher than that for β-phase over the entire composition as shown in Fig. 3.5(a), the free energy of a mixture of α- and β-phases is always higher than that of α-phase. In other words, a mixture of two phases can't exist. When the free energy of α-phase opens upwards, two α-phases with composition $(p' + q')$ or $(p'' + q'')$ have free energy higher than that of a single α-phase with composition $(G_\alpha)_c$ (Fig. 3.5(b)). A single α-phase is always at equilibrium. To summarize,

I) When the free energy of α-phase is lower than that of β-phase over the entire composition, α-phase is the equilibrium phase over the entire composition (Fig. 3.1(a)).

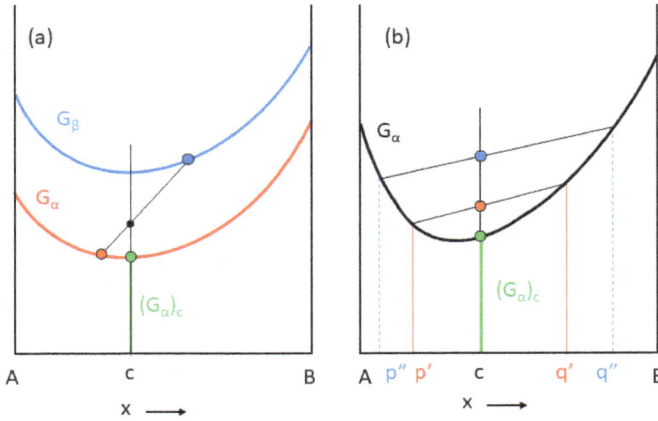

Fig. 3.5 (a) Free energy of α-phase is lower over the entire composition. (b) Free energy of α-phase opens upward.

II) By contrast in the case shown in Fig. 3.2, composition of $x_\bullet$ is divided into three regions, where p and q are points of contact on the common tangential line for α- and β-phases.

① $0 < x_\bullet < p$: Free energy of α-phase is lower than that of β-phase. That is, α-phase is the equilibrium phase.

② $q < x_\bullet < 1$: Free energy of β-phase is lower than that of α-phase. That is, β-phase is the equilibrium phase.

③ $p < x_\bullet < q$: $(\alpha)_q$ is at equilibrium with $(\beta)_q$.

When intermediate phases are present in addition to the primary α- and β-phases in the binary A-B system, the free energy-composition curves are schematically shown in Fig. 3.6.

3.1.2   Simple phase diagram and free energy

First draw the free-energy/composition curves at different temperatures, then apply the law of common tangential line.

a. Isomorphous system (complete solid solution)

Only α-solid solution and liquid (L) are present. Free-energy/composition curves at different temperatures and the resultant phase diagram are shown qualitatively in Fig. 3.7.

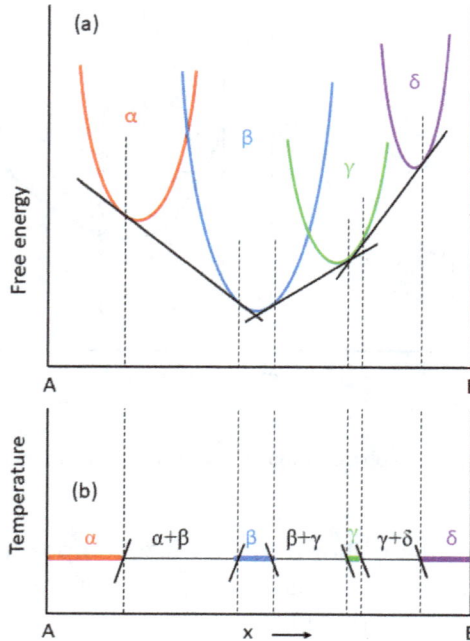

Fig. 3.6   Free energies of intermediate phases.

b. Eutectic alloy

Three phases, liquid (L), α-solid solution and β-solid solution are present. Free-energy/composition curves at different temperatures and the resultant phase diagram are shown qualitatively in Fig. 3.8.

Exercise. $T = T_4$ is the eutectic temperature. In the eutectic reaction, when three phases, i.e., $L + \alpha + \beta$, are coexistent, the temperature ($T_e = T_4$) and compositions of the α- and β-phases are determined uniquely. Discuss the significance of $f = 0$.

Exercise. Draw free-energy/composition curves for a peritectic reaction.

Exercise. A congruent alloy must have either a maximum or a minimum in its phase diagram (see Fig. 1.43). Discuss the reason for this in terms of the free-energy/composition curves.

Exercise. Discuss the exercise in Fig. 1.44 in terms of free-energy/composition curves.

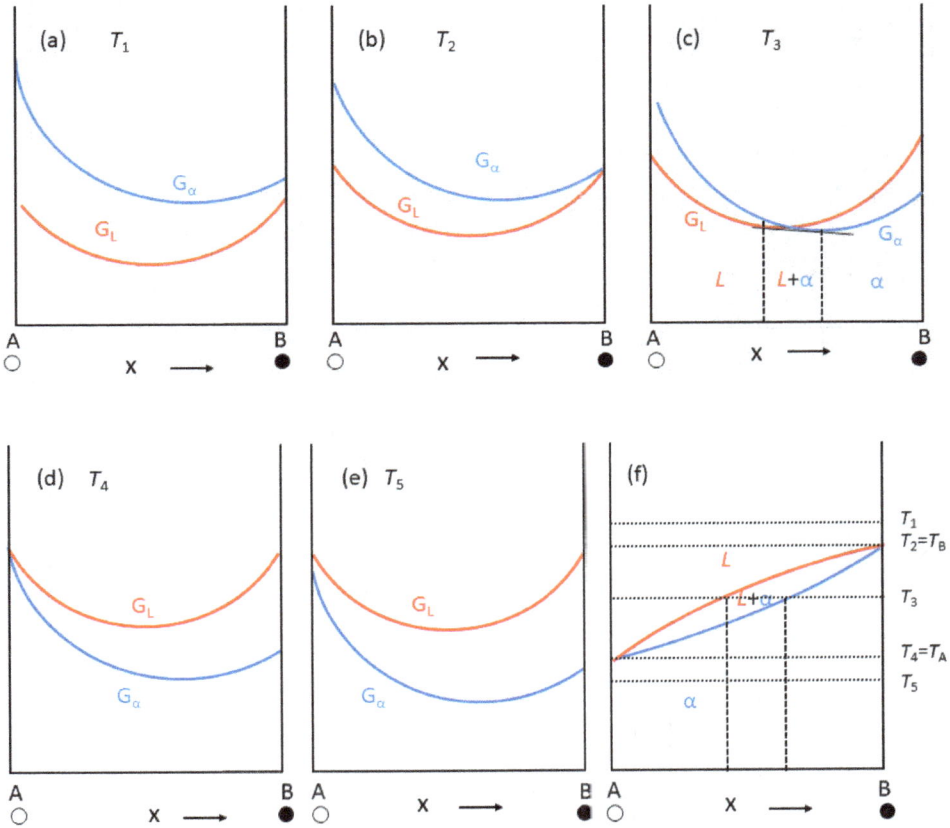

Fig. 3.7 (a)–(e) Free-energy/composition curves of an isomorphous system. (a) T_1, (b) T_2, (c) T_3, (4) $T_4 = T_A$, (5) T_5, (f) Phase diagram obtained using (a)–(e).

3.1.3 Free energy of substitutional solid solution

In the foregoing discussion it is assumed that free-energy/composition curve of substitutional solid solution opens upwards. In what follows let us draw concrete free-energy/composition curves.

Free energy treated here is one under a constant pressure and at a constant temperature (absolute temperature T), i.e. the Gibbs' free energy G which is given by

$$G = H - TS \tag{3.3}$$

where H and S are cohesive enthalpy and entropy at absolute temperature T, respectively. Enthalpy H is given by $H = \bar{E} + pV$, where E is internal

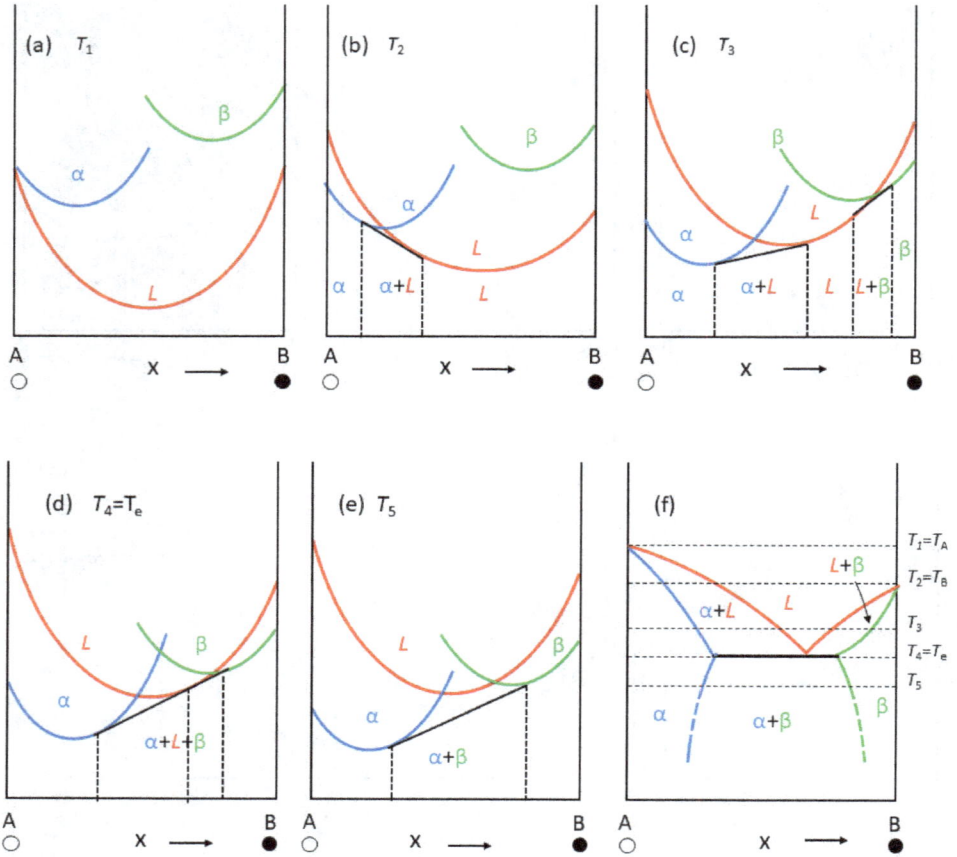

Fig. 3.8 (a)–(e) Free-energy/composition curves of a eutectic system. (a) T_1, (b) $T_2 = T_B$, (c) T_3, (4) $T_4 = T_e$ (eutectic temperature), (5) T_5, (f) Phase diagram obtained using (a)–(e).

(a)   (b)

Fig. 1.43(a)(b) (repeat).

energy, p is pressure and V is volume. In the case of metallic alloys, only solids and liquids under a constant pressure are treated, so that pV can be neglected. Therefore, Eq. (3.3) is rewritten as $G \fallingdotseq E - TS$, however, following suit let us write it as $G = H - TS$.

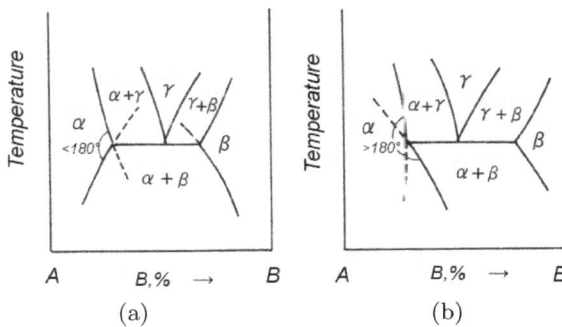

Fig. 1.44(a)(b) (repeat).

Now let us discuss the free energy of individual phases. For the sake of simplicity, let us concentrate on a substitutional solid solution; an interstitial solid solution is not treated.

a. Cohesive enthalpy of a substitutional solid solution

Figure 3.9(a) shows a two-dimensional model of a binary substitutional solid solution. Lattice points are regularly arranged, each of which is occupied by either A($\circ$) or B($\bullet$) atoms. Figure 3.9(b) shows schematically a model where cavities correspond to lattice points and atoms are trapped inside the cavities. The deeper the cavity, the bigger the binding energy.

Figure 3.9(c) illustrates interatomic potential as a function of atomic distance (r), where E_{ij} (i, j = A, B) is the binding energy. In the binary system A($\circ$) atoms and B($\bullet$) atoms are present, so that three binding pairs, A-A, B-B and A-B, are to be considered as follows:

Binding energy of A-A ($\circ$-$\circ$) pair E_{AA} ($=E_{\circ\circ}$)
Binding energy of B-B ($\bullet$-$\bullet$) pair E_{BB} ($=E_{\bullet\bullet}$)
Binding energy of A-B ($\circ$-$\bullet$) pair E_{AB} ($=E_{\circ\bullet}$)

Sum of these binding energies are cohesive enthalpy. It is to be noted that binding energy is negative, i.e., $E_{ij} < 0$.

In order to obtain summation, the numbers of A-A ($\circ$-$\circ$), B-B ($\bullet$-$\bullet$) and A-B ($\circ$-$\bullet$) binding pairs must be calculated. The numbers of the binding pairs are calculated as follows.

Suppose that the number of A($\circ$) atoms is $N_{\circ}$, and that of B($\bullet$) atoms $N_{\bullet}$.

$$N_{\circ} + N_{\bullet} = N \tag{3.4a}$$

where N is the total number of atoms in the alloy system.

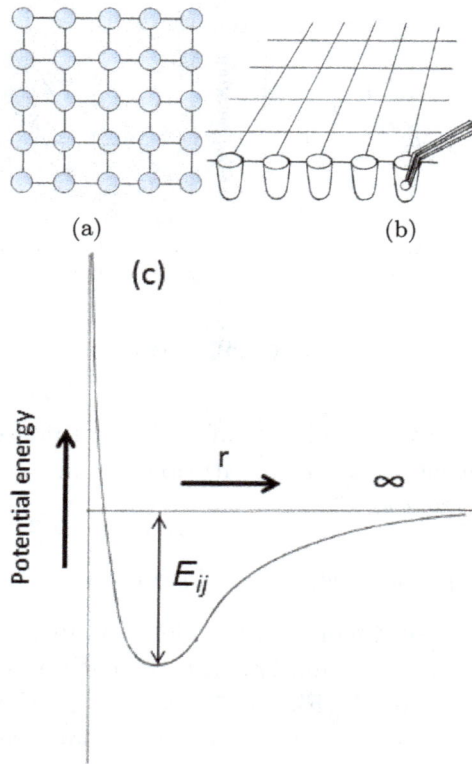

Fig. 3.9 (a)(b) Two-dimensional model of a binary substitutional solid solution. (c) Interatomic potential as a function of atomic distance (r).

Atomic fractions of A($\circ$) and B($\bullet$) atoms $x_{\circ}$ and $x_{\bullet}$ are given by

$$x_{\circ} = N_{\circ}/N \tag{3.4b}$$

$$x_{\bullet} = N_{\bullet}/N \tag{3.4c}$$

So that

$$x_{\circ} + x_{\bullet} = 1 \tag{3.5}$$

Let the coordinate number (the number of the nearest atoms, four in the case of Fig. 3.9(a)) be z, then probability at which one of the z nearest atoms around any one of the A($\circ$) atoms is occupied by A($\circ$) atom is $x_{\circ} = N_{\circ}/N$. Therefore, probability of forming A-A ($\circ$-$\circ$) pair around one of the A($\circ$) atoms is z $(N_{\circ}/N)$. The total number of A($\circ$) in the overall crystal is $N_{\circ}$, thus the total number of A-A (o-o) pair [oo] is given by[2]

$$[oo] = z \times \frac{N_o}{N} \times N_o \times \frac{1}{2} = \frac{zN_o^2}{2N} \qquad (3.6a)$$

Similarly,

$$[\bullet\bullet] = \frac{zN_\bullet^2}{2N} \qquad (3.6b)$$

$$[o\bullet] = \frac{1}{2}\left(\frac{zN_o}{N} \times N_\bullet + \frac{zN_\bullet}{N} \times N_o\right) = \frac{zN_oN_\bullet}{N} \qquad (3.6c)$$

Therefore, the cohesive enthalpy $^oH^\alpha$ of α-solid solution at absolute zero of temperature (0 K) is given by

$$^oH^\alpha = [oo] \times E_{oo} + [\bullet\bullet] \times E_{\bullet\bullet} + [o\bullet] \times E_{o\bullet}$$
$$= \frac{zN}{2}\left[E_{oo}(1-x_\bullet) + E_{\bullet\bullet}x_\bullet + 2\left(E_{o\bullet} - \frac{E_{oo}+E_{\bullet\bullet}}{2}\right)(1-x_\bullet)x_\bullet\right]$$
$$= {}^oH_o^\alpha(1-x_\bullet) + {}^oH_\bullet^\alpha x_\bullet + \Omega_{o\bullet}^\alpha(1-x_\bullet)x_\bullet \qquad (3.7)$$

where

$$^oH_o^\alpha = \frac{N \cdot z \cdot E_{oo}}{2} \qquad (3.8a)$$

$$^oH_\bullet^\alpha = \frac{N \cdot z \cdot E_{\bullet\bullet}}{2} \qquad (3.8b)$$

$$\Omega_{o\bullet}^\alpha = Nz\left(E_{o\bullet} - \frac{E_{oo}+E_{\bullet\bullet}}{2}\right) \qquad (3.8c)$$

$^oH_o^\alpha$ is the cohesive enthalpy of pure A (o) metal, $^oH_\bullet^\alpha$ is the cohesive enthalpy of pure B ($\bullet$) metal and $\Omega_{o\bullet}^\alpha$ (hereinafter denoted by Ω without subscripts o and $\bullet$ since it is obvious that the interaction is between o and $\bullet$) is interaction parameter.

The interaction parameters can be categorized into three cases (Fig. 3.10);

(1) $\Omega^\alpha = 0$
(2) $\Omega^\alpha > 0$
(3) $\Omega^\alpha < 0$

(1) This is for the ideal solution, where A(o) and B($\bullet$) are mixed without discrimination.

---

[2]The reason for the existence of 1/2 is that the same [oo] pair is counted twice.

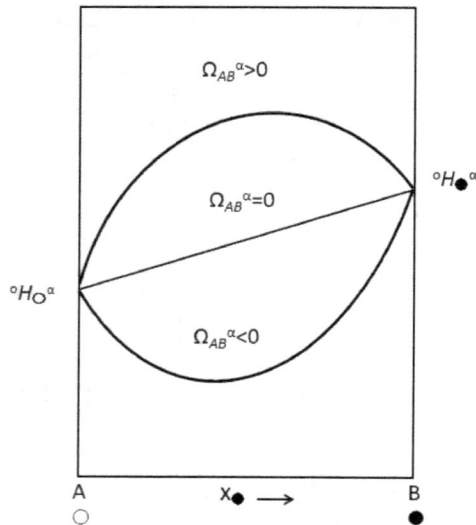

Fig. 3.10   Cohesive enthalpy $^\circ H^\alpha$ versus $x_\bullet$ for $\Omega > 0$, $\Omega = 0$ and $\Omega < 0$.

(2) In this case, the binding energy of A-B (○-●) pair is larger than the average value of A-A (○-○) and B-B (●-●) pairs.   As a result, the A-A (○-○) pair and B-B (●-●) pair are preferred to the A-B (○-●) pair, leading to two-phase separation.

(3) This is the reverse of case (2). That is, the average value of A-A (○-○) pair and B-B (●-●) pair is larger than that of A-B (○-●) pair, so that the A-B (○-●) pair is preferred to the A-A (○-○) and B-B (●-●) pairs, leading to ordering.

b. Entropy of configuration in a substitutional solid solution

Entropy is one of the most difficult concepts in thermodynamics. Entropy prescribes the most probable state. Thus, the law of the maximum entropy is nothing but a law to say that what should happen indeed happens.

Let us consider the case where 25 lattice points are occupied with both A and B atoms in Fig. 3.9(a). Concretely, a total of 25 atoms (either A or B atoms; $N_A + N_B = 25$) are distributed onto 25 lattice points, and then let us count the number (W) of how they are distributed.

First suppose that $N_A = 1$ and $N_B = 24$. In other words, one A atom is distributed to any one of the 25 lattice points, so that the number of distribution $W_1 = 25$. Next suppose that $N_A = 2$ and $N_B = 23$. One lattice point out of 25 is already occupied by A atom, so that the number of distributing the second A atom to the remaining 24 lattice points is 24.

Therefore, the number of ways to distribute two A atoms is given by

$$W_2 = 25 \times 24/2$$

The reason why it is divided by 2 is as follows: Out of two A atoms, which one A atom occupies the lattice point first in the position-taking game?

| A atom 1 | Place 1 | Place 2 |
| A atom 2 | Place 2 | Place 1 |

The number of distributing the third A atom 3 ($N_A = 3$, $N_B = 22$) is given by

$$W_3 = 25 \times 24 \times 23/(3 \times 2)$$

(3×2) in the denominator is present because there are six ways in the position-taking game for three A atoms.

A atom 1	Place1	Place1	Place 2	Place 3	Place 2	Place 3
A atom 2	Place 2	Place 3	Place 1	place1	Place 3	Place 2
A atom 3	Place 3	Place 2	Place 3	Place 2	Place 1	Place 1

First, which one, out of three A atoms, is the first place does not matter, so that the number of distributing is 3 (shaded). For each of the remaining two A atoms are placed as the second or the third, in two ways. Therefore, the total number of distribution is $3 \times 2 = 6$.

Exercise. When there are four atoms (i.e., $N_A = 4$, $N_B = 21$) the number of distribution, W_4, is given by $W_4 = 25 \times 24 \times 23 \times 22/(4 \times 3 \times 2)$. Prove this.

To summarize, the number of ways in which $N_\circ$ of A($\circ$) atoms and $N_\bullet$ of B($\bullet$) atoms are distributed (W) is given by

$$\begin{aligned}
W &= \frac{(N-1)(N-2)\cdots(N-N_\circ)}{N_A!} \\
&= \frac{N(N-1)(N-2)\cdots(N_\bullet+1)}{N_A} \\
&= \frac{N!}{N_\circ! N_\bullet!}
\end{aligned} \tag{3.9}$$

Here,

$$N! = N(N-1)(N-2)(N-3)\cdots 1$$

According to statistical thermodynamics, entropy ^0S is defined as

$$^0S = k\ln W \tag{3.10}$$

where k is Boltzmann constant ($= 1.38\times10^{-23}$ J·K^{-1} $= 8.62\times10^{-5}$ eV·K^{-1}). Substituting Eq. (3.9) gives

$$^0S = k(\ln N! - \ln N_{\circ}! - \ln N_{\bullet}!) \tag{3.10a}$$

When N is very large, the Stirling formula

$$\ln N! = N\ln N - N$$

is applied. In summary,

$$
\begin{aligned}
^0S &= k(N\ln N - N - N_{\circ}\ln N_{\circ} + N_{\circ} - N_{\bullet}\ln N_{\bullet} + N_{\bullet})\\
&= -Nk[(x_{\bullet}\ln x_{\bullet} + (1-x_{\bullet})\ln(1-x_{\bullet})]
\end{aligned}
\tag{3.10b}
$$

For 1 mole

$$^0S^{\alpha}_{\circ\bullet} = -R[(x_{\bullet}\ln x_{\bullet} + (1-x_{\bullet})\ln(1-x_{\bullet}))]$$

where R is the gas constant ($= 1.986$ cal/K· mol).

As is shown in Fig. 3.11 entropy shows a maximum at $x_{\bullet} = 0.5$.

c. Free energy of a substitutional solid solution

Free energy G at temperature T is given by

$$G = H - TS \tag{3.11}$$

where H and S are enthalpy and entropy at absolute temperature T, respectively.

$$H = {}^0H + \int_0^T C_p dT \tag{3.12a}$$

$$S = {}^0S + \int_0^T \frac{C_p}{T} dT \tag{3.12b}$$

where ^0H and ^0S are enthalpy and entropy at absolute zero of temperature

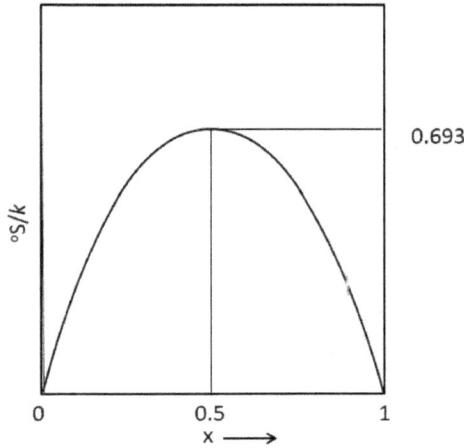

Fig. 3.11 Entropy of configuration.

$T = 0$ K, respectively, and C_p is the specific heat at constant pressure (isopiestic specific heat). Therefore, the free energy G^α of a solid solution at a finite temperature T is given by

$$
\begin{aligned}
G^\alpha = {}&^0H^\alpha_\circ(1 - x_\bullet) + {}^0H^\alpha_\bullet \cdot x_\bullet + \Omega^\alpha(1 - x_\bullet)x_\bullet \\
&+ RT(x_\bullet \ln x_\bullet + (1 - x_\bullet)\ln(1 - x_\bullet)) \\
&+ \int_0^T C^\alpha_p dT - T \int_0^T \frac{C^\alpha_p}{T} dT
\end{aligned}
\tag{3.13}
$$

Generally speaking, C_p varies approximately linearly with $x_\bullet$, and thus can be neglected when only the concentration dependence of G^α is considered.

Thus,

$$
\begin{aligned}
G^\alpha = {}&^0G^\alpha_\circ(1 - x_\bullet) + {}^0G^\alpha_\bullet \cdot x_\bullet + \Omega^\alpha(1 - x_\bullet)x_\bullet \\
&+ RT(x_\bullet \ln x_\bullet + (1 - x_\bullet)\ln(1 - x_\bullet))
\end{aligned}
\tag{3.14}
$$

where $^0G^\alpha_\circ$ and $^0G^\alpha_\bullet$ are the free energies of pure A ($\circ$) and B ($\bullet$) metals at absolute zero of temperature $T = 0$ K, respectively.

Plotting G^α for

$$\Omega^\alpha = 0$$
$$\Omega^\alpha > 0$$
$$\Omega^\alpha < 0$$

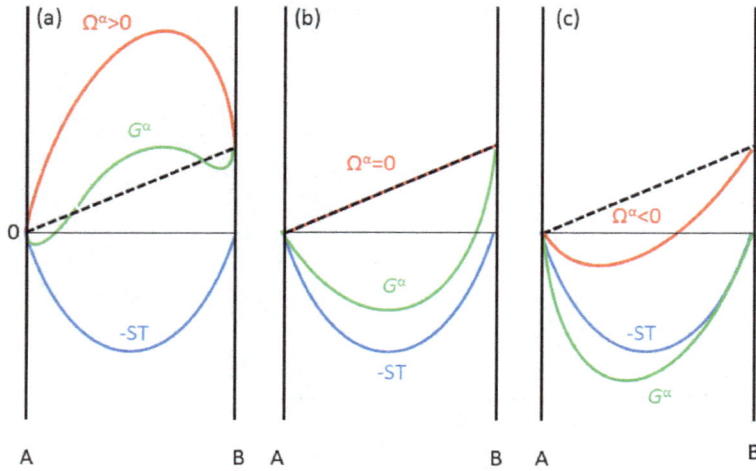

Fig. 3.12   Free energy/composition curves for $\Omega > 0$, $\Omega = 0$ and $\Omega < 0$.

gives Fig. 3.12. Except for the case for $\Omega^\alpha > 0$, the free energy of an α-solid solution shows a minimum at around $x_\bullet \fallingdotseq 0.5$.

Now, let us discuss in more detail the case where $\Omega^\alpha > 0$. In this case, two-phase separation takes place. Figure 1.38 illustrates schematically a model of the phase diagram of this type. Let us make that $G_A^\alpha = G_B^\alpha = G$ in Eq. (3.14) for the sake of simplicity.

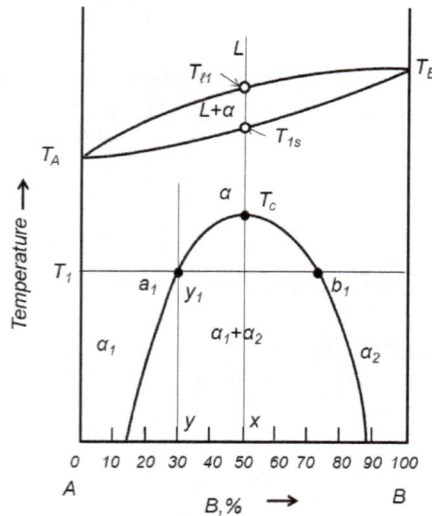

Fig. 1.38 (repeat).

$$G^\alpha = G + \Omega^\alpha(1 - x_\bullet)x_\bullet + RT\left[x_\bullet\ln x_\bullet + x_\circ\ln x_\circ\right]$$
$$= G + \Omega^\alpha(1 - x_\bullet)x_\bullet + RT\left[x_\bullet\ln x_\bullet + (1 - x_\bullet)\ln(1 - x_\bullet)\right] \qquad (3.14')$$

G is plotted as functions of $x_\bullet$ for different parameter RT/Ω in (Fig. 3.13). As is shown for $RT/\Omega = 0.4$ by a broken line, α-phase decomposes into two phases. Differentiating G by $x_\bullet$,

$$\frac{dG^\alpha}{dx_\bullet} = \Omega^\alpha - 2\Omega^\alpha x_\bullet + RT[\ln x_\bullet - \ln(1 - x_\bullet)] \qquad (3.15)$$

Solubility limits correspond to the two minima which are given by

$$\frac{dG^\alpha}{dx_\bullet} = 0$$

Solving

$$\frac{dG^\alpha}{dx_\bullet} = \Omega^\alpha(1 - x_\bullet) + RT[\ln x_\bullet - \ln(1 - x_\bullet)] = 0$$

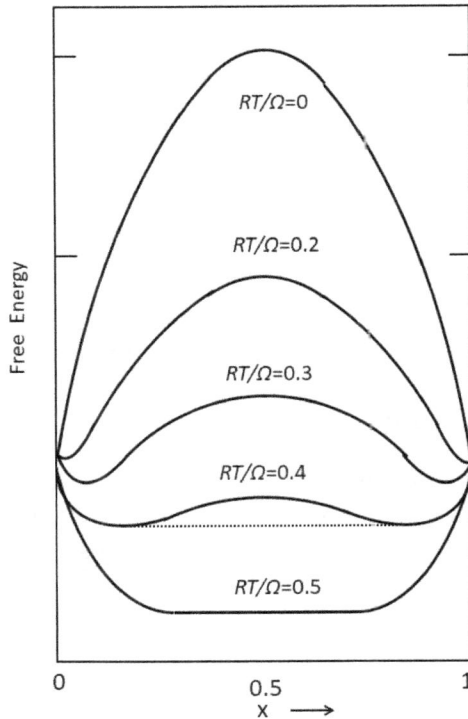

Fig. 3.13 Free energy in a system where two phase separation takes place.

the two minima are obtained. When two solutions are obtained, they correspond to the two ends of a tie line (a_1 and b_1 in Fig. 1.38). At the critical temperature T_c, the two solutions converse to one solution.

In order to obtain T_c, let us consider the points of inflection instead of minima themselves, since both minima and the points of inflection converse at T_c.

Thus, the solution for

$$\frac{d^2 G^\alpha}{dx_\bullet^2} = 0$$

has only one solution. That is, solving

$$\frac{d^2 G^\alpha}{dx_\bullet^2} = -2\Omega^\alpha + RT \left[\frac{1}{x_\bullet} - \frac{1}{1-x_\bullet} \right] = 0$$

solutions for the point of inflection are given by

$$\frac{1}{x_\bullet} - \frac{1}{1-x_\bullet} = \frac{2\Omega^\alpha}{RT}$$

Putting $RT/\Omega = A$,

$$\frac{1}{x_\bullet} - \frac{1}{1-x_\bullet} = \frac{1}{A}$$

leading to

$$x_\bullet^2 - x_\bullet + A = 0$$

This quadratic equation has only one solution at T_c, thus from discriminant $1 - 4A = 0$. That is, $RT_c/\Omega = A = 1/4$. In other words, $T_c = \Omega/2R$.

3.1.4   Free energy of a liquid

The density of a liquid is more or less similar to that of a solid, suggesting that the configurations of atoms in a liquid and a solid can be regarded as similar to each other at least in the short range. This is a semi-crystal model of a liquid. The free energy of a liquid can be attained by modifying slightly that of a solid solution.

a. Cohesive energy of liquid

Cohesive energy of a liquid is assumed to be higher than that of a solid solution by the latent heat.

$$^0H^L_{\circ} \cong {}^0H^{\alpha}_{\circ} + L_{\circ}$$
$$^0H^L_{\bullet} \cong {}^0H^{\alpha}_{\bullet} + L_{\bullet}$$

(3.16)

where $L_{\circ}$ and $L_{\bullet}$ are the latent heats of fusion of pure metal A($\circ$) and pure metals B($\bullet$), respectively.

b. Entropy of liquid

Entropy of a liquid is assumed to be higher than that of a solid solution by the latent heat of fusion.

$$\int_0^T \frac{C_p^L}{T} dT = \int_0^T \frac{C_p^{\alpha}}{T} dT + \frac{L}{T_m}$$

(3.17)

where T_m is the absolute melting point.

Now, let us use Richards relation, that is, $L \fallingdotseq RT_m$, then Eqs. (3.16) and (3.17) are given by

$$^0H^L_{\circ} \cong {}^0H^{\alpha}_{\circ} + RT_{\circ}$$
$$^0H^L_{\bullet} \cong {}^0H^{\alpha}_{\bullet} + RT_{\bullet}$$
$$\int_0^T \frac{C_p^L}{T} dT \cong \int_0^T \frac{C_p^{\alpha}}{T} dT + R$$

(3.18)

By analogy to Eq. (3.13),

$$G^L = \left({}^0H^{\alpha}_{\circ} + RT_{\circ}\right)(1 - x_{\bullet}) + \left({}^0H^{\alpha}_{\bullet} + RT_{\bullet}\right) x_{\bullet} + \Omega^L(1 - x_{\bullet})x_{\bullet}$$
$$+ RT[x_{\bullet}\ln x_{\bullet} + (1 - x_{\bullet})\ln(1 - x_{\bullet})]$$
$$+ \int_0^T C_p^{\alpha} dT - T\int_0^T \frac{C_p^{\alpha}}{T} dT - RT$$

(3.19)

As in the case of G^{α},

$$G^L(c) = \left({}^0G^{\alpha}_{\circ} + RT_A\right)(1 - x_{\bullet}) + \left({}^0G^{\alpha}_{\bullet} + RT_B\right) x_{\bullet} + \Omega^L(1 - x_{\bullet})x_{\bullet}$$
$$+ RT[x_{\bullet}\ln x_{\bullet} + (1 - x_{\bullet})\ln(1 - x_{\bullet})] - RT$$

(3.20)

where Ω^L_{AB} is the interaction parameter in a liquid.

From Eqs. (3.19)–(3.20),

$$
\begin{array}{ll}
G^{\alpha}{}_{\circ}, G^{\alpha}{}_{\bullet}, T_A, T_B & \text{known} \\
\Omega^{\alpha}_{AB}, \Omega^{L}_{AB} & \text{important}
\end{array}
$$

When the above six values are known, the free energies of a solid and a liquid at different temperatures can be calculated as functions of concentration $x_{\bullet}$.

c. Derivation of fundamental binary phase diagrams through thermodynamics

For the sake of simplicity let us make $G^{\alpha}_A = G^{\alpha}_B = 0$. Also let us make $x_{\bullet} = c$ in Eqs. (3.19) and (3.20), then

$$G^{\alpha}(c) = \Omega^{\alpha}(1-c)c + RT[c\ln c + (1-c)\ln(1-c)] \tag{3.19'}$$

$$G^{L}(c) = RT_A(1-c) + RcT_B + \Omega^{L} + (1-c)c - RT + RT[c\ln c + (1-c)\ln(1-c)] \tag{3.20'}$$

Exercise. Constitute a binary phase diagram for $T_A = 900$, $T_B = 1300$, $\Omega^{L} = 0$ and $\Omega^{\alpha} = -2000$ cal/mol.

Hint. An isomorphous solid solution with the liquidus and solidus (both convex).

Exercise. Constitute a binary phase diagram $T_A = 900$, $T_B = 1300$, $\Omega^{L} = 2000$ cal/mol and $\Omega^{\alpha} = 6000$ cal/mol.

Hint. Eutectic system.

3.2 Thermodynamics of Nucleation

3.2.1 Spinodal decomposition

Let us discuss in more detail how the two-phase decomposition described in Fig. 3.13 proceeds. As already mentioned, there are minima and points of inflection in the free-energy/composition curve below T_c (see Fig. 3.14). The minima correspond to the solubility limits a_1 and b_1 in Fig. 2.17 (law of common tangential line). Existence of points of inflection has an important significance.

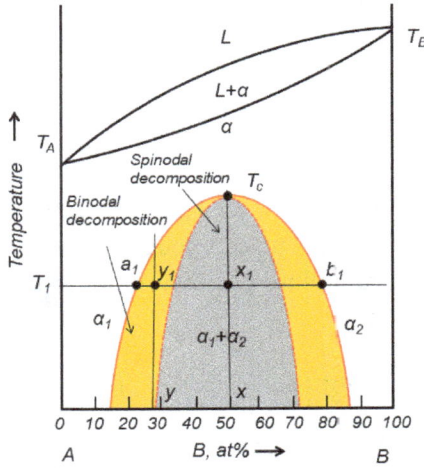

Fig. 2.17 (repeat).

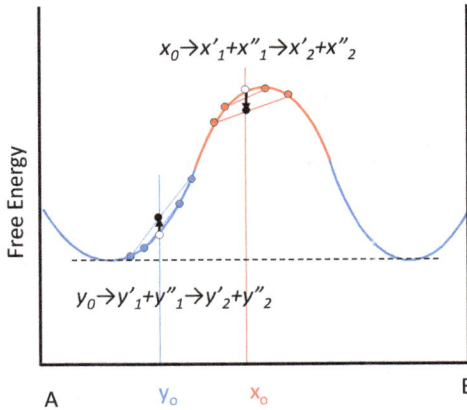

Fig. 3.14 Variation in free energy in spinodal decomposition (x_0) and binodal decomposition (y_0).

Suppose that an α-solid solution is quenched from a temperature above T_c, followed by aging at T_1 in Fig. 2.17. The α-solid solution is supercooled and on subsequent aging decomposes as $\alpha \rightarrow (\alpha_1)_{a_1} + (\alpha_2)_{b_1}$.

However, at the very beginning of decomposition, the equilibrium composition of α_1 is unlikely to be a_1, and similarly the composition of α_2 is unlikely to be b_1. It is reasonable to assume that the supercooled solid solution decomposes to two phases whose compositions are closer to the original gross composition of the supercooled solid solution. In this process, what happens inside and outside the points of inflection are definitely different.

Inside the points of inflection (for instance, x_0) the decomposition proceeds as follows:

$$x_0 \rightarrow x_1' + x_1'' \rightarrow x_2' + x_2''$$

Here, the free energy of the product (mixture of α_1 and α_2) decreases monotonously from the original value x_o. In other words, decomposition takes place spontaneously without an incubation period (Fig. 2.14).

By contrast, outside the points of inflection (for instance, y_0) the decomposition proceeds as follows:

$$y_0 \rightarrow y_1' + y_1'' \rightarrow y_1' + y_2''$$

Here, the free energy of the product (mixture of α_1 and α_2) increases monotonously from the original value y_o. In other words, in order for decomposition to proceed, it is necessary to overcome a hill of energy. Thus, the incubation period appears (Fig. 2.14). The decompostion inside the points of inflection is referred to as spinocal decomposition and that outside to binodal decomposition (Fig. 2.17).

3.2.2   Homogeneous nucleation

Another reason why an incubation period appears is the formation of interfaces accompanied by phase transformation. An interface has a higher energy than the matrix; this energy is referred to as an interfacial energy and expressed in terms of energy per unit area, like erg/cm^2 or mJ/m^2.

Let us consider solidification, that is, transformation from a liquid to a solid as an example. Now suppose that a solid particle with radius r is formed inside a liquid. This (small) particle is referred to as an embryo. Volume of the embryo is $(4/3)\,\pi r^3$. Therefore, assuming that the change in free energy per unit volume associated with transformation is ΔG_v, the gain in total free energy is $(4/3)\,\pi r^3 \Delta G_v$. On the other hand, an interface with an area of $4\pi r^2$ is formed between the solid and liquid (solid-liquid interface). If the interfacial energy has γ per unit area, this results in an increase in energy by $4\pi r^2 \gamma$.

Therefore, the net change in free energy ΔG is given by

$$\Delta G = \frac{4}{3}\pi \mathrm{r}^3 \Delta G_v + 4\pi r^2 \gamma \tag{3.21}$$

The first term is negative, while the second term is positive. Thus, ΔG shows a maximum (ΔG^*) at a certain critical value of r^* (Fig. 3.15(a)). r^*

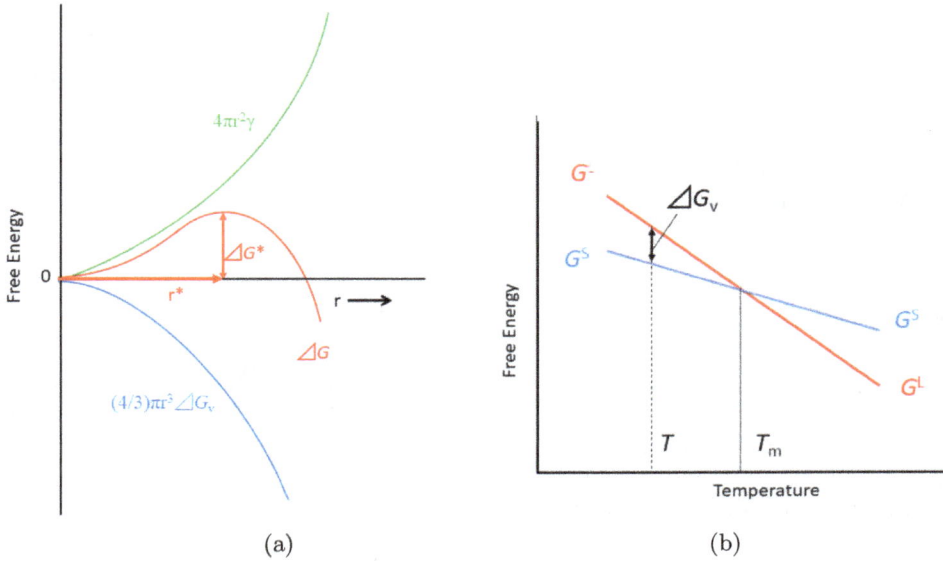

Fig. 3.15 (a) Variation in free energy in homogeneous nucleation. (b) G^L and G^S versus temperature.

and ΔG^* are obtained by differentiating ΔG in Eq. (3.21) by r, and setting

$$\frac{d\Delta G}{dr} = 4\pi r^2 \Delta G_v + 8\pi r\gamma = 0$$
$$r^* = -\frac{2\gamma}{\Delta G_v} \tag{3.22}$$

$$\Delta G^* = \frac{16\pi\gamma^3}{3\Delta G_v^2} \tag{3.23}$$

The significance of Fig. 3.15(a) is that when the radius of an embryo is smaller than the critical radius r^*, it can't overcome the barrier ΔG^* and disappears. Only those embroyes which happen to exceed the critical radius r^* succeed in growing further. Such a successful embryo is referred to as a critical nucleus, and ΔG^* is referred to as the critical nucleation energy.

The critical radius is inversely proportional to ΔG_v. Let us consider ΔG_v in more detail. Free energies of a liquid and a solid are plotted as functions of temperature in Fig. 3.15(b). ΔG_v is the difference in the two curves at a temperature T below the melting point T_m. That is,

$$\Delta G_v = G^S - G^L = (H^S - H^L) - T(S^S - S^L)$$
$$\Delta T = T_m - T \tag{3.24}$$

where $\Delta T = T_m - T$ is the degree of supercooling.

At the melting point T_m

$$G^S = G^L$$

so that

$$\Delta G_v = G^S - G^L = (H^S - H^L) - T_m(S^S - S^L) = 0$$
$$\therefore S^S - S^L = \frac{\Delta H}{T_m} \tag{3.25}$$

Substituting this into Eq. (3.24)

$$\Delta G_v = \Delta H - \frac{T \cdot \Delta H}{T_m} \tag{3.26}$$

Therefore, Eq. (3.22) gives

$$r^* = -\frac{2\gamma T_m}{\Delta H \cdot \Delta T} \tag{3.27}$$

$$\Delta G^* = \frac{16\pi\gamma^3}{3\Delta G_v^2} = \frac{16\pi\gamma^3 T_m^2}{3\Delta H^2 \cdot \Delta T^2} \tag{3.28}$$

Equations (3.27) and (3.28) indicate that with increasing supercooling (ΔT) the critical radius decreases, so does the critical nucleation energy (ΔG^*).

3.2.3 Heterogeneous nucleation

In practice, solidification starts with a degree of supercooling which is much smaller than expected from homogeneous nucleation. This is accounted for by the nucleation taking place at the surface of an inclusion and/or wall of a crucible.

Suppose that a solid is formed at the surface of inclusion or crucible as shown in Fig. 3.16. The change in energy is given by

$$\Delta G = V_s(G^S - G^L) + [S_{sm}(\gamma_{sm} - \gamma_{lm}) + \gamma_{sl} \cdot S_{sl}] \tag{3.29}$$

where V_s is the volume of a solid nuclei, S_{sm} is the area of interface between

Liquid

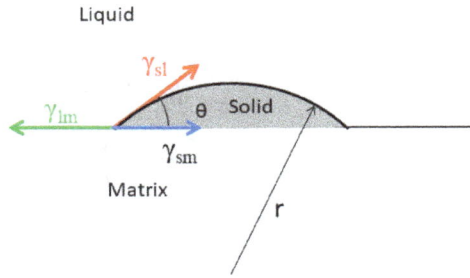

Fig. 3.16 Young's relation.

solid and matrix, and S_{sl} the area of interface between solid and liquid. The first and second terms correspond to those in Eq. (3.21). γ_{sm}, γ_{lm}, and γ_{sl} are the interfacial energies (per unit area) between a solid-matrix, a liquid-matrix and a solid-liquid, respectively. Subscripts l, m and s stand for liquid, matrix and solid, respectively.

Exercise. Derive Eq. (3.29).

Here, among γ_{sm}, γ_{lm} and γ_{sl} the following equation holds due to the balance of the forces

$$\gamma_{lm} = \gamma_{sm} + \gamma_{sl} \cos\theta \tag{3.30}$$

Equation (3.30) is referred to as Young equation, and θ is referred to as the contact angle or wetting angle. Taking this relation into consideration, a treatment analogue to that in the preceding section leads to

$$\Delta G^* = \frac{4\pi\gamma^3 T_m^2}{3\Delta H^2 \cdot \Delta T^2}(2 + \cos\theta)(1 - \cos\theta)^2 \tag{3.31}$$

Exercise. Derive Eq. (3.31).

Hint. As is shown in Fig. 3.17, the volume (V_a) and the surface area (S_a) of a sphere truncated at a position at a distance $\boldsymbol{a}$ away from the origin are given by

$$V_a = \pi\left(\frac{2}{3}r^3 - r^2 a + \frac{a^3}{3}\right)$$
$$S_a = 2\pi r(r - a)$$

The significance of this is that the formation energy of critical nuclei in heterogeneous nucleation is equal to that in homogeneous nucleation (Eq. (3.28)) multiplied by a factor $(2 + \cos\theta)(1 - \cos\theta)^2/4$.

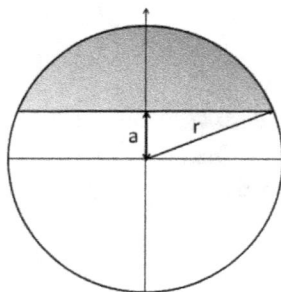

Fig. 3.17

Since $1 \geq \cos\theta \geq 0$, ΔG^* in the heterogeneous nucleation is smaller than that in the homogeneous nucleation, so that heterogeneous nucleation is much easier than the homogeneous nucleation.

Exercise. Show that, when $\theta = 0°$

$$(2 + \cos\theta)(1 - \cos\theta)^2/4 = 0$$

at $\theta = 180°$

$$(2 + \cos\theta)(1 - \cos\theta)^2/4 = 1.$$

For the respective cases, draw equivalents to Fig. 3.16 and discuss their physical significance.

3.2.4   Precipitation of intermediate phases

In Sec. 2.1.3, it was mentioned that precipitation proceeds in sequence and that GP zones or metastable phases can be formed prior to the equilibrium phases. This can be accounted for as follow.

Suppose that, in Fig. 3.18, α-phase with composition c $((\alpha)_c)$ decomposes into $(\alpha)_{a3}$ and $(\beta)_{b3}$. That is,

$$\alpha_c \rightarrow \alpha_3 + \beta$$

Suppose also that metastable β_1 and β_2 phases are present, the compositions of which are closer to that of α-phase than that of β-phase. The gain in energy due to the decomposition

$$\alpha_c \rightarrow \alpha_3 + \beta$$

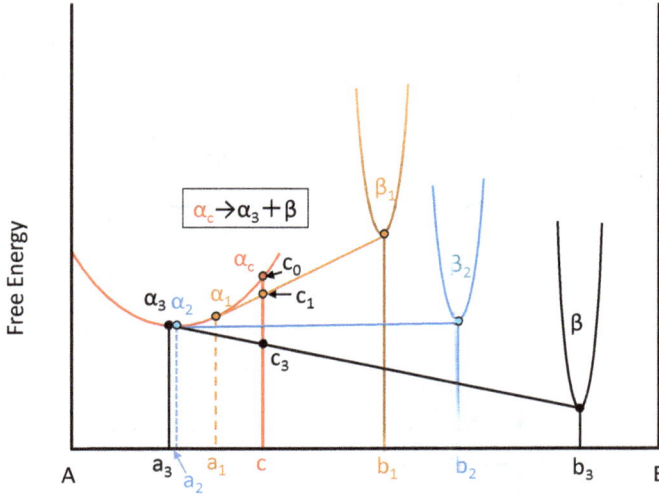

Fig. 3.18 Variation in free energy/composition curves during precipitation of interme-diate phases.

is $c_0 c_3$. However, in order for this decomposition to take place, $(\alpha)_c$ must decompose into a_3 and b_3 whose compositions are far away from c.

By contrast, when β_1 precipitates as a metastable phase, the energy gain is only $c_0 c_1$, which is smaller than $c_0 c_3$. However, the change in composition is only a_1, b_1, much smaller than a_3 and b_3. On top of this, when β_1 is coherent with matrix like G.P. zone, the interfacial energy due to an interface accompanied by precipitation is smaller. Therefore, the first step of precipitation is β_1 metastable phase (GP zone), then β_2 phase precipitates as an intermediate phase. This is thermodynamic reasoning for formation of metastable intermediate phases in precipitation sequence.

3.2.5 Martensitic transformation

The most important feature of a martensitic transformation is that it is a diffusionless transformation. In other words, the composition during the transformation remains unchanged. Let us discuss a martensitic transformation in Fe-Ni binary system (Fig. 3.19). When an alloy with composition c is quenched from γ to α-phases, it passes T_1, T_2 and T_3. The free-energy/composition curves at respective temperatures are shown in Fig. 3.19. At T_1 the energy of $(\gamma)_c$ is higher than that of $(\alpha)_c$, so that the martensitic transformation has not yet started. At T_2, $(\gamma)_c = (\alpha)_c$, so that T_2 is thermodynamic transformation temperature (corresponding to T_0

in Fig. 2.28). Even so, due to the formation of an interface accompanied by the formation of martensite, the transformation has not yet started. External stress can induce martensitic transformation, the upper limit of temperature at which this strain-induced martensitic transformation can be induced is referred to as T_d. T_d can't exceed T_0.

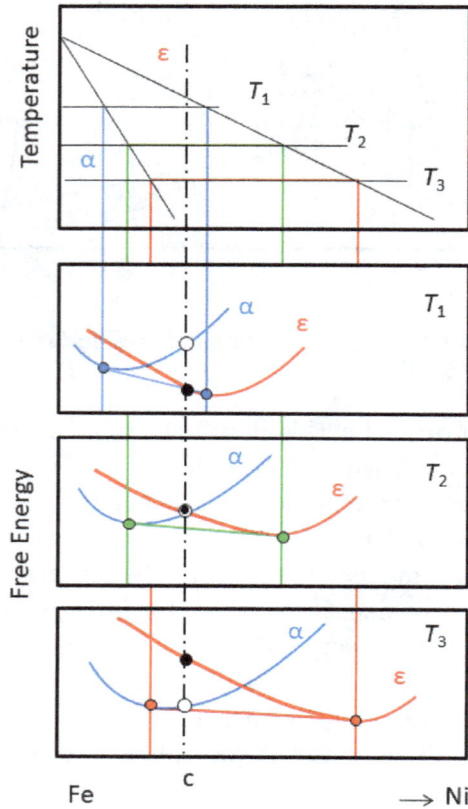

Fig. 3.19　Free energy/composition curves in martensitic transformation.

Chapter 4

Ternary Phase Diagram

4.1 Fundamentals of Ternary Phase Diagrams

Multi-phase alloys are used in practical service. For instance, ternary alloys include important alloys such as stainless steels (Fe-Cr-Ni alloy). Ternary phase diagrams are far more complex than binary phase diagrams. This chapter deals with the fundamentals of ternary phase diagrams.

4.1.1 How to represent — The Gibbs triangle

A ternary alloy consists of three elements, A, B and C. In order to represent the chemical composition, the Gibbs equilateral triangle with three vertices of A, B and C is used in the triangular coordinate (Fig. 4.1). Three sides of the equilateral triangle correspond to the component binary phase diagrams of A-B, B-C and C-A alloys, respectively. Let the perpendicular bisector of the side be h, and perpendicular bisectors from a point $\mathbf{x}$ inside the triangle to sides BC, CA and AB be p, q and r, respectively, then p corresponds to concentration of A (hereinafter denoted by %A), q to %B,[1] r to %C.[1]

Exercise. Prove that $p + q + r = h = 100\%$

In Fig. 4.2 the compositions at points o, p and q which lie on a line connecting vertex C and a point $\mathbf{r}$ on the side AB are given by

$$\frac{oy}{oC} = \frac{py'}{pC} = \frac{qy''}{qC} = \frac{ry'''}{rC}$$

$$\frac{oz}{oC} = \frac{pz'}{pC} = \frac{qz''}{qC} = \frac{qz'''}{rC}$$

$$\therefore \frac{oy}{oz} = \frac{py'}{pz'} = \frac{qy''}{qz''} = \frac{ry'''}{rz'''} = \frac{Ar}{rB}$$

---

[1]Similarly, %B and %C stand for the concentrations of B and C, respectively.

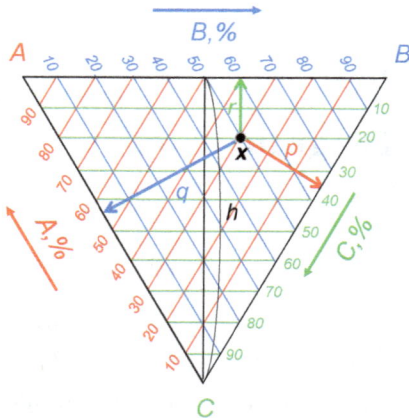

Fig. 4.1 The Gibbs triangle. Concentrations (1).

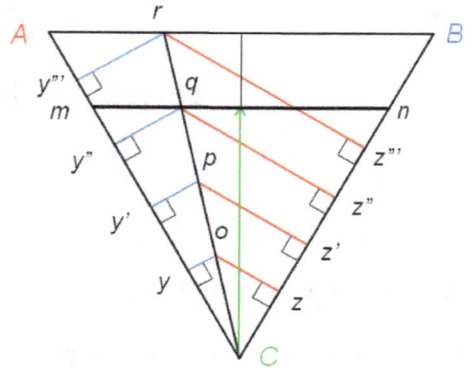

Fig. 4.2 The Gibbs triangle. Concentrations (2). Along **mn**, %C is constant. Along $C\boldsymbol{r}$, %B:%C is constant.

That is, the ratio %B:%C is constant. By contrast, for points along **mn** which is parallel to side AB, %C is constant (%C = constant). On moving along **mn**, only the ratio of %A:%B changes.

4.1.2   Tie-lines and lever principle

a. Tie-lines

Let us suppose that two alloys **a** and **b** belonging to one ternary system are mixed together with a ratio of $m{:}n$, then the composition of the product alloy (**c**) lies on a line connecting m and n (Fig. 4.3). Point of **c** is obtained by a lever principle as follows:

$$m{:}n = \overline{bc}{:}\overline{ac}$$

Exercise. Show that composition **c** is given by the lever principle in Fig. 4.3(b).

b. Tie-triangle

The composition of an alloy (P) formed by mixing three alloys all of which belonging to the same ternary system (L, R and S) lies within a triangle whose vertices are L, R and S, respectively (Fig. 4.4(a)). Let the amounts

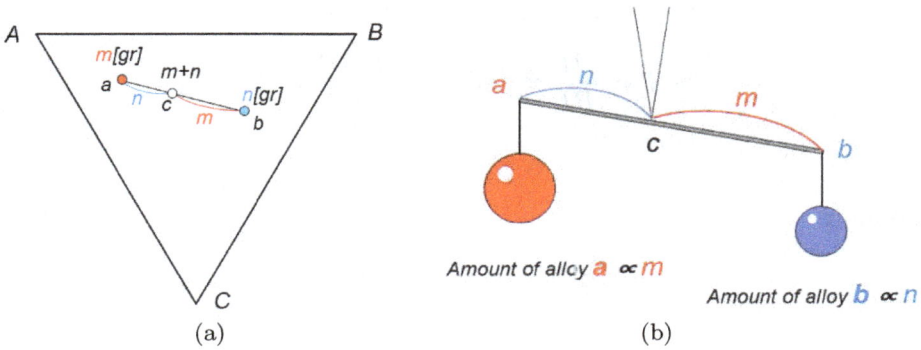

Fig. 4.3 Lever principle (1).

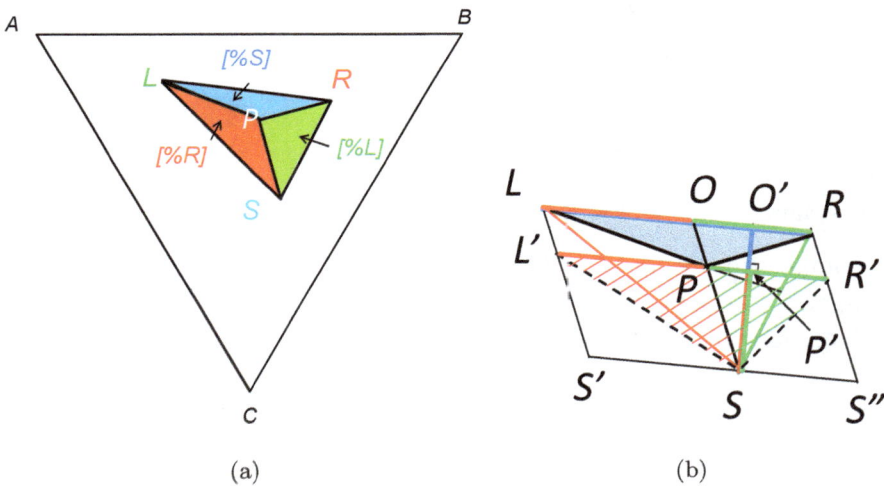

Fig. 4.4 Lever principle (2).

of S, R and L be denoted by $[\%S]$, $[\%R]$, $[\%L]$, then $[\%S]:[\%R]:[\%L] = \Delta PRL : \Delta SPL : \Delta SRP$.

Proof. In Fig. 4.4(b), an alloy is formed by mixing alloys S and O. Here, the lever principle described in Sec. 4.1.2a states

$$[\%S] = \frac{PO}{OS} \times 100$$

$$[\%O] = \frac{SP}{OS} \times 100$$

Next, the alloy O is formed by mixing alloys L and R. Then,

$$[\%R] = \frac{OL}{RL} \times [\%O] = \frac{OL}{RL}\frac{SP}{OS} \times 100$$

$$[\%L] = \frac{RO}{RL} \times [\%O] = \frac{RO}{RL}\frac{SP}{OS} \times 100$$

Therefore,

$$
\begin{array}{ccc}
[\%S]: & [\%R]: & [\%L] \\[6pt]
= \dfrac{PO}{OS}: & \dfrac{OL}{RL}\dfrac{SP}{OS}: & \dfrac{RO}{RL}\dfrac{SP}{OS} \\[10pt]
= PO: & \dfrac{OL \cdot SP}{RL}: & \dfrac{RO \cdot SP}{RL} \\[10pt]
= PO \cdot RL: & OL \cdot SP: & RO \cdot SP \\[6pt]
= (P'O' \times RL): & (OL \times SP'): & (RO \times SP') \\[6pt]
= \Delta PRL: & \Delta SPL': & \Delta SR'P \\[6pt]
= \Delta PRL: & \Delta SPL: & \Delta SRP
\end{array}
$$

c. Tie-square

The composition of an alloy (R) formed by mixing four alloys all of which belonging to the same ternary system $(V_1, V_2, V_3$ and $V_4)$ lies within a square whose corners are V_1, V_2, V_3 and V_4 (Fig. 4.5).

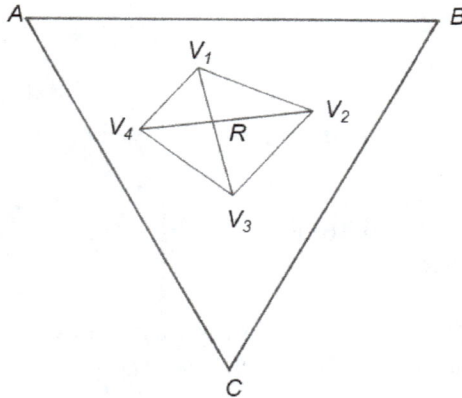

Fig. 4.5   Lever principle (3).

Let the amounts of V_1, V_2, V_3 and V_4 be $[\%V_1]$, $[\%V_2]$, $[\%V_3]$ and $[\%V_4]$, then

$$[\%V_1] : \qquad [\%V_2] : \qquad [\%V_3] : \qquad [\%V_4]$$
$$= \Delta V_2 V_3 V_4 : \quad \Delta V_1 V_3 V_4 : \quad \Delta V_1 V_2 V_4 : \quad \Delta V_1 V_2 V_3$$

4.1.3 Phase rule in the ternary system

The degree of freedom in a ternary system is given by

$$f = c + 1 - p = 4 - p$$
$$(\because c = 3)$$

Therefore, when four phases are coexistent, the system is invariant.

Table 4.1

Number of phase (p)	4	3	2	1
Degree of freedom (f)	0	1	2	3

4.1.4 Space diagram and sectional diagram

a. Space diagram

The compositions are expressed in a triangular coordinate and temperature is expressed by a coordinate vertical to the triangular coordinate. In other words, a ternary phase diagram is expressed by a triangular cylinder.

The simplest ternary system, that is, an isomorphous (complete) solid solution is shown in its space diagram in Fig. 4.6(a). What corresponds to the solidus and liquidus lines in the binary phase diagram are the solidus and liquidus surfaces. Figure 4.6(b) shows a projection of contours of the liquidus surface.

b. Sectional diagram

While a space diagram may appear easy to understand at first sight, it is difficult to visualize a complex phase diagram by a space diagram. This is the reason why a cross section of a space diagram is necessary. Such a section is referred to as a sectional diagram. When a space diagram is sectioned horizontally, the section is referred to as an isothermal sectional diagram (isotherm), and when sectioned vertically the section is referred to as a vertical sectional diagram.

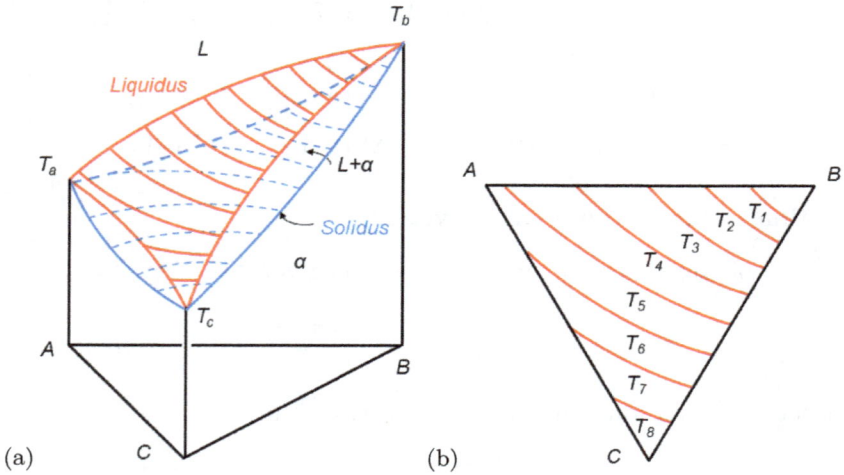

Fig. 4.6 (a) Solidus and liquidus of an isomorphous solid solution. (b) Projections of contours of the liquidus of an isomorphous solid solution.

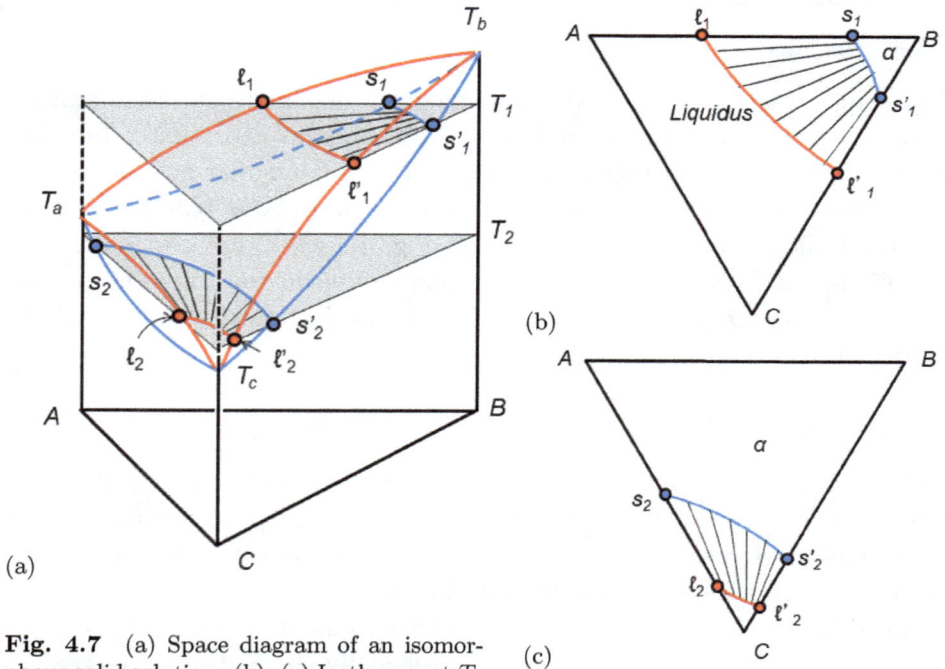

Fig. 4.7 (a) Space diagram of an isomorphous solid solution. (b), (c) Isotherms at T_1 and T_2.

Isotherms of a complete ternary solid solution are shown in Fig. 4.7(a). At T_1, intersections of an isotherm with the liquidus $\ell_1\ell_2$ and the solidus s_1s_2 appear (Figs. 4.7(b)(c)). Straight lines inside the $(L + \alpha)$ two-phase field in Figs. 4.7(b)(c) are the tie lines between liquid (L) and solid (α). These lines are to be determined experimentally.

A typical vertical section is shown in Fig. 4.8. Let us consider xB and xy as sections. First, let us consider a section along xB (Fig. 4.8(a)). Draw a vertical line passing through x in the A-C binary system, then its intersections with the solidus x_s and the liquidus x_ℓ are determined. Since B is a pure metal, the liquidus and solidus converge at T_b. The vertical section is obtained by connecting x_ℓ, x_s and T_b as shown in Fig. 4.8(c).

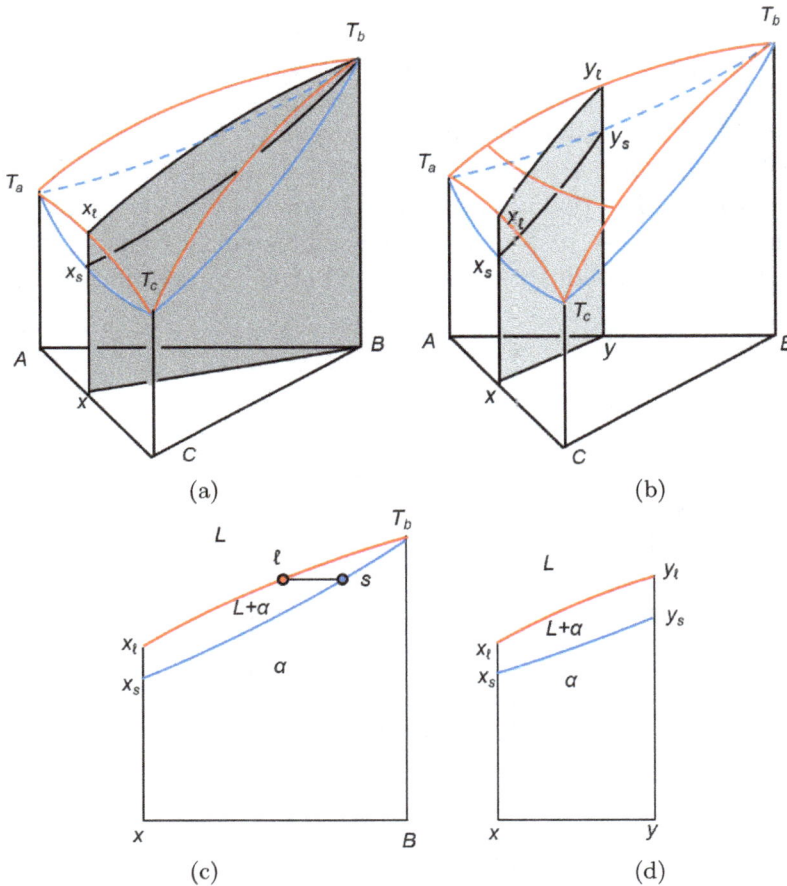

Fig. 4.8 (a)(b) Vertical sections of an isomorphous solid solution in space diagram. (c) Vertical sections along xB (see (a)) and (d) along xY (see (b)).

Next, let us consider a vertical section passing through xY (Fig. 4.8(b)). Draw a vertical line passing through Y in the A-B binary system, then y_ℓ and y_s are determined. Connecting x_ℓ-y_ℓ and x_s-y_s, a vertical sectional diagram is obtained as shown in Fig. 4.8(d). A vertical sectional diagram is useful to overlook a wide temperature range. However, ℓs which is a horizontal line in $(L + \alpha)$ field in Fig. 4.8(c) is not a tie line. A tie line must be drawn in a two-phase field in the isotherm (Figs. 4.7(b)(c)).

4.2   Relatively Simple Ternary Systems

4.2.1   Solidification of an isomorphous (complete) ternary solid solution

Let us consider the solidification process of an alloy with composition X. Figure 4.9 shows a space diagram and Figs. 4.10(a), (b), (c) and (d) show isotherms at T_1, T_2, T_3 and T_4. As can be seen from the space diagram, at T_1 the liquid X contacts the liquidus surface at ℓ_1. The counterpart α-solid solution has composition α_1, and L_1-α_1 is a tie line passing through X at T_1 (Fig. 4.10(a)). The tie lines at T_2 and T_3 are $\alpha_2\ell_2$ and $\alpha_3\ell_3$, respectively. At T_4 the liquid phase of X reaches the solidus surface and α_4 on the tie line now lies on the solidus surface $\ell_4\alpha_4$. That is, the solidification finishes.

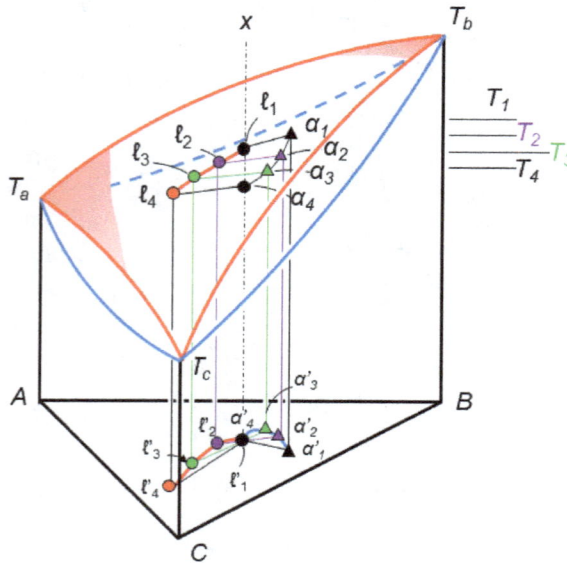

Fig. 4.9   Freezing process of an isomorphous solid solution. Freezing starts at T_1 and finishes at T_4.

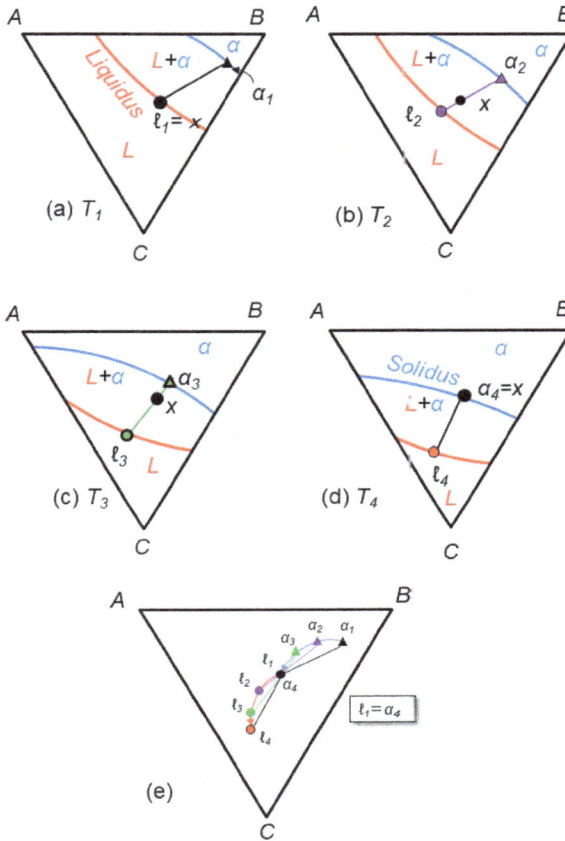

Fig. 4.10 Freezing process of an isomorphous solid solution. Freezing starts at T_1 and finishes at T_4. (e) indicates projections of tie lines during freezing. For instance $\alpha_1 l_1$ is a tie line at T_1.

Compositions of α-solid solution and the liquid phase change along the traces of $\alpha_1 \rightarrow \alpha_2 \rightarrow \alpha_3 \rightarrow \alpha_4$ and $l_1 \rightarrow l_2 \rightarrow l_3 \rightarrow l_4$, respectively in Fig. 4.10(e).

4.2.2 Simple eutectic system

Figures 4.11(a), (b) and (c) show a simple eutectic system. Figures 4.11(a) and (b) show the space diagram with the component binary systems A-B and B-C shaded, respectively. Figure 4.11(c) shows an opening-up of the space diagram ((a) and (b)) with a projection of the phase boundaries viewed downward added (hereinafter referred to as an opened-up phase diagram). Binary A-B and B-C systems are eutectic systems, but binary C-A system is an isomorphous solid solution. Shaded fields in Fig. 4.11(a) and

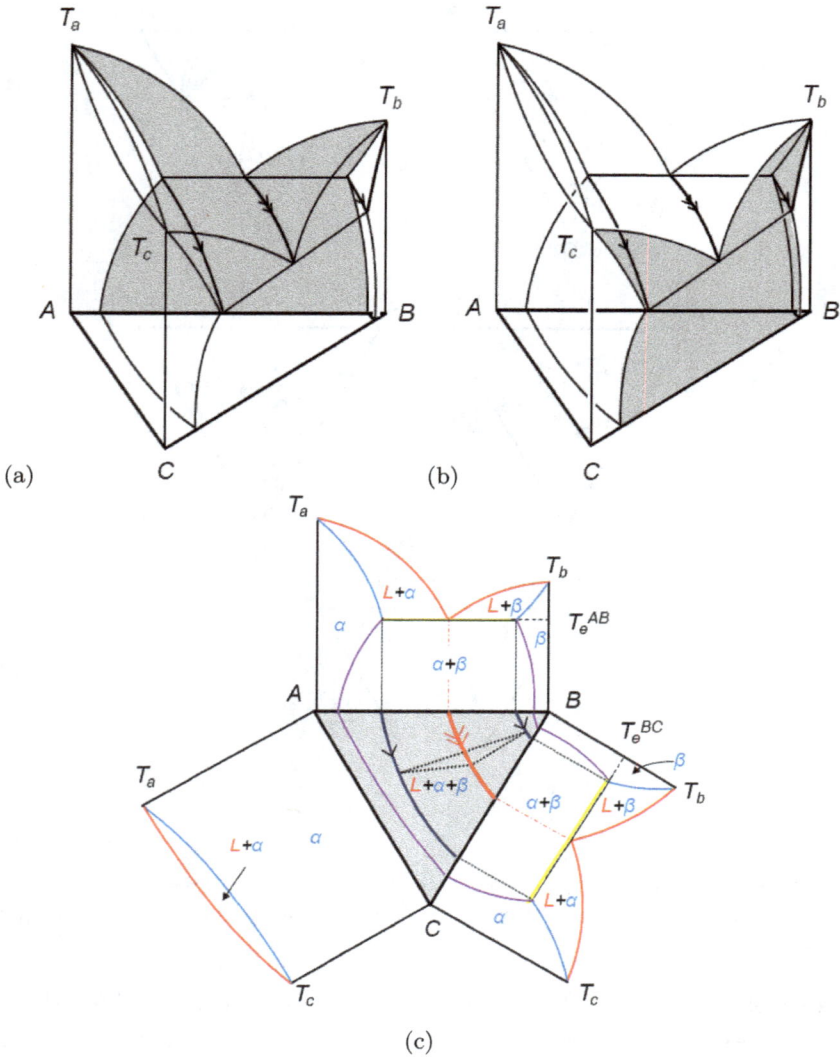

Fig. 4.11 (a) Eutectic reaction in a ternary system. Shaded area indicates the A-B binary eutectic system. (b) Eutectic type. Shaded area indicates the B-C binary eutectic system. Eutectic temperature of the AB system T_e^{AB} is higher than that of the B-C system T_e^{BC}. (c) Projection of an opened-up space diagram.

Fig. 4.11(b) correspond to A-B binary eutectic system and C-B binary eutectic system, respectively. The eutectic point in A-B binary system (T_e^{AB}) is higher than that in B-C binary system (T_e^{BC}). In the ternary system the eutectic temperature descends from T_e^{AB} to T_e^{BC} along a bold line indicated by a double-headed arrow (Fig. 4.11(c)).

The key point here is that, while the degree of freedom in the three-phase field $(L + \alpha + \beta)$ is zero $(f = 0)$ in a binary system, in a ternary system $f = 1$. In other words, the triangle of three-phase $(L + \alpha + \beta)$ field has a thickness with two roofs and one broad base (Fig. 4.12).

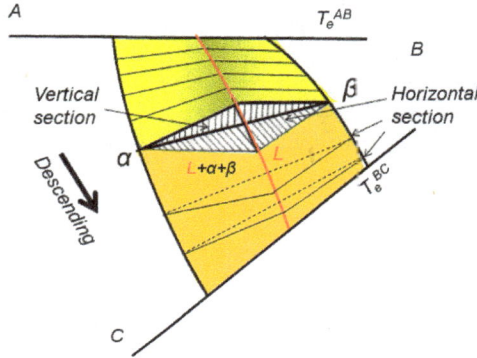

Fig. 4.12 Three-phase triangle $(L + \alpha + \beta)$ in a eutectic system. T_e^{AB} is the eutectic temperature of the A-B binary eutectic system and T_e^{BC} is the eutectic temperature of the B-C binary eutectic system. T_e^{AB} is higher than T_e^{EC}.

Now, let us decompose surfaces showing the phase boundaries (hereinafter denoted by PBS; phase-boundary surfaces) shown in Figs. 4.11(a) and (b) from above (high temperature) layer by layer. The first PBS is the liquidus (Fig. 4.13(a) and Fig. 4.14(a)). The liquidus consists of two surfaces, i.e., $T_a\text{-}T_e^{AB}\text{-}T_e^{BC}\text{-}T_c$ and $T_e^{AB}\text{-}T_e^{BC}\text{-}T_b$. The next PBS's are two solidus surfaces $\left(T_a\text{-}s_\alpha^{AB}\text{-}s_\alpha^{BC}\text{-}T_c \text{ and } T_b\text{-}s_\beta^{BC}\text{-}s_\beta^{AB}\right)$ and the roofs of the $(L + \alpha + \beta)$ three-phase triangle $\left(s_\beta^{AB}\text{-}e^{AB}\text{-}e^{AC}\text{-}s_\beta^{BC} \text{ and } e^{AB}\text{-}s_\beta^{AB}\text{-}s_\beta^{BC}\text{-}e^{BC}\right)$ (Fig. 4.13(b) and Fig. 4.14(b)).

The last PBS's are the broad base of the $(L + \alpha + \beta)$ three-phase triangle $\left(s_\alpha^{AB}\text{-}s_\beta^{AB}\text{-}s_\beta^{BC}\text{-}s_\alpha^{BC}\right)$ and two solvus surfaces, $\left(s_\alpha^{AB}\text{-}{}^os_\alpha^{AB}\text{-}{}^os_\alpha^{BC}\text{-}s_\alpha^{BC}\right)$ and $\left(s_\alpha^{AB}\text{-}{}^os_\alpha^{AB}\text{-}{}^os_\alpha^{BC}\text{-}s_\alpha^{BC}\right)$, which represent $(\alpha + \beta)$ field and α, β single phases (Fig. 4.13(c) and Fig. 4.14(c)).

This is summarized in the following Table 4.2.

Drawing isotherms and vertical sections from Figs. 4.13 and 4.14 facilitates understanding a ternary phase diagram. In the following the detailed procedures of these will be presented.

Table 4.2

	A-C binary system		Pure B metal	
Layer 1	T_a-T_e^{AB}-T_e^{BC}-T_c Liquidus		T_e^{AB}-T_e^{BC}-T_b liquidus	
Layer 2	T_a-s_a^{AB}-s_a^{BC}-T_c Solidus surface	$(L+\alpha+\beta)$ Roofs of 3-phase triangle (convex)	$(L+\alpha+\beta)$ Roofs of 3-phase triangle (convex)	T_b-s_β^{BC}-s_β^{AB} Solidus surface
Layer 3	α phase's solvus	$(L+\alpha+\beta)$ Lower broad face of 3-phase triangle		β phase's solvus

a. Isotherms

First prepare an opened-up diagram shown in Fig. 4.11(c) but with the central ternary triangle blank (Fig. 4.15(a)). Let liquidus in binary A-B and B-C systems drawn in red, solidus lines in navy blue and solvus lines in purple.

Draw the line corresponding to $T_1(> T_e^{BC}, T_e^{AB})$ in the binary A-B, B-C and C-A systems, then in the binary C-A system the T_1 line does not intersect a phase boundary. By contrast in the A-B binary system, the T_1 line intersects α solidus at $\alpha_1(\bullet)$, liquidus line at two points $((\ell_\alpha, \ell_\beta)(\bullet))$, and β solidus at $(\beta_1(\bullet))$. Draw the vertical lines from these four points down to room temperature and let the points of intersections be denoted by $(\underline{\alpha_1}, \underline{\ell_\alpha}, \underline{\ell_\beta}, \underline{\beta_1})$. Draw phase boundaries obtained from the binary A-B system along line AB. Similar procedure on the B-C binary system leads to $\underline{\alpha'_1}, \underline{\ell'_\alpha}, \underline{\ell'_\beta}$ and $\underline{\beta'_1}$. Here, $(')$ indicates the B-C binary system.

Next, draw short lines from $(\underline{\alpha_1}, \underline{\ell_\alpha}, \underline{\ell_\beta}, \underline{\beta_1})$ and $(\underline{\alpha'_1}, \underline{\ell'_\alpha}, \underline{\ell'_\beta}, \underline{\beta'_1})$ toward the inside of the A-B-C ternary triangle (Fig. 4.15(b)). Now it is obvious that the phase boundaries inside the A-B-C ternary system is something like Fig. 4.15(c).

Now let us consider $T_2(T_e^{AB} > T_2 > T_e^{BC})$ (Fig. 4.16(a)). Procedure similar to T_1 leads to Fig. 4.16(b). Two $\bullet$'s and two $\bullet$'s can be connected without difficulty. Two $\bullet$'s in the B-C binary system, which are not connected with short lines from the A-B binary system, connect themselves to form a loop (Fig. 4.16(c)). Then, judging from the surrounding coexisting phases, $(L + \alpha + \beta)$ ternary field will be drawn as shown in Fig. 4.16(d). Drawing tie lines in two-phase fields $(L + \alpha)$, $(L + \beta)$ and $(\alpha + \beta)$, isotherm at T_2 is obtained (Fig. 4.16(e)).

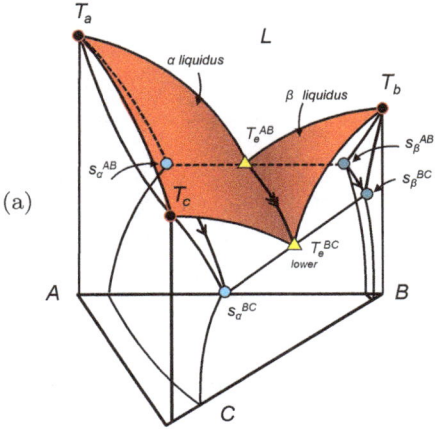

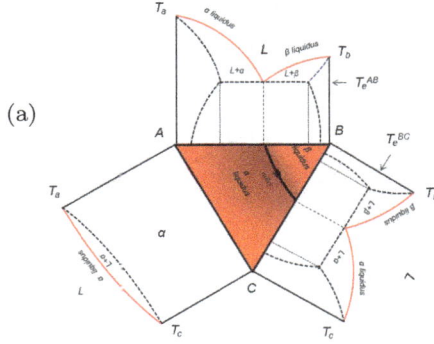

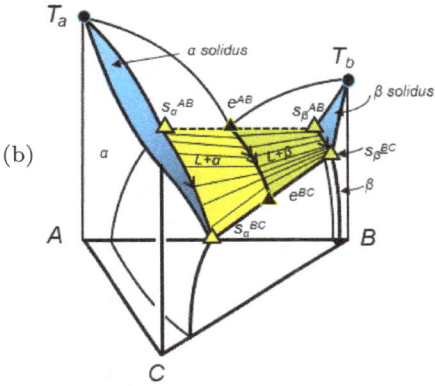

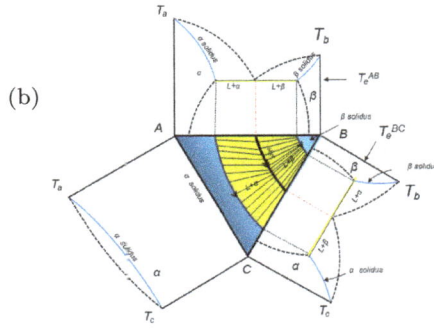

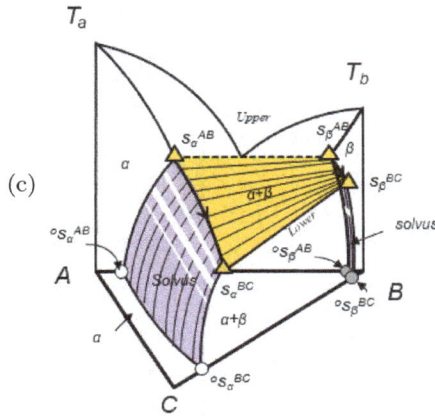

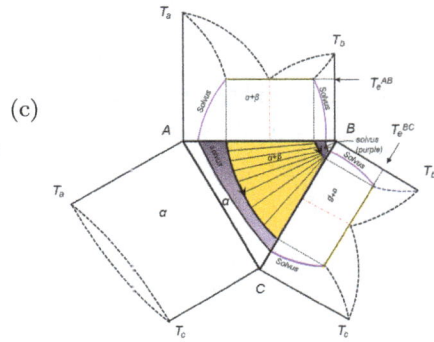

Fig. 4.13 (a) 1st PBS, (b) 2nd PBS, (c) 3rd PBS.

Fig. 4.14 (a) 1st PBS, (b) 2nd PBS, (c) 3rd PBS.

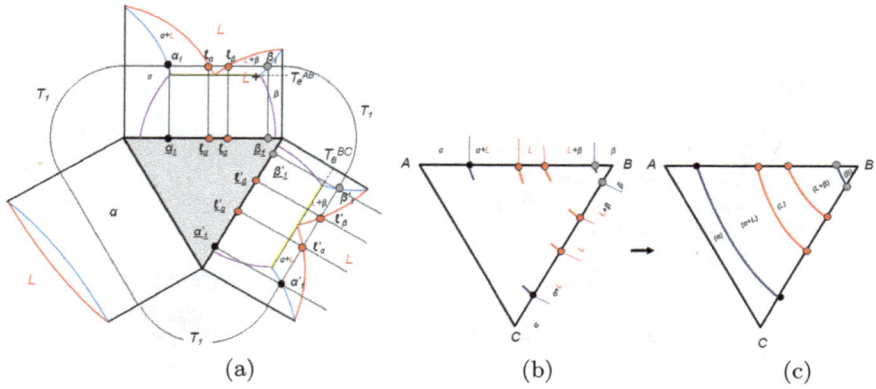

Fig. 4.15   Drawing an isotherm at T_1. (a) Procedure ①. (b), (c) Procedure ②.

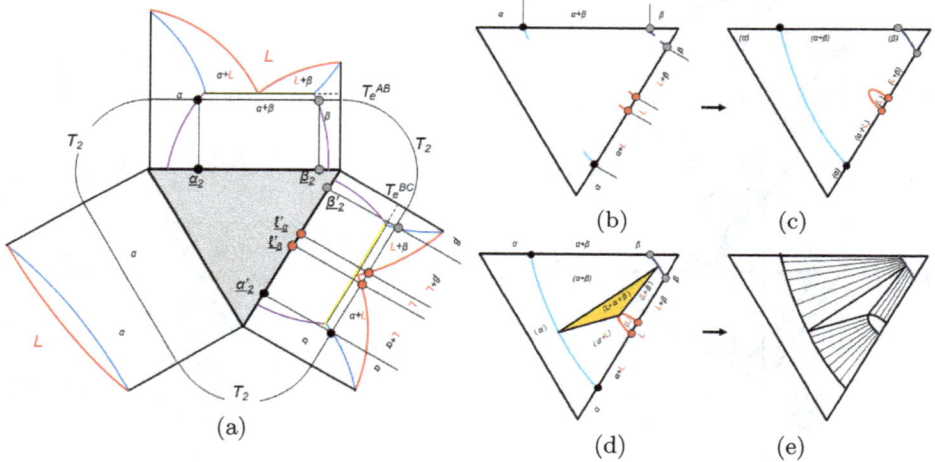

Fig. 4.16   Drawing an isotherm at T_2. (a) Procedure ①. (b), (c) Procedure ②. (d), (e) Procedure ③.

Similar procedure at $T_3 (< T_e^{AB}, T_e^{BC})$ leads to Fig. 4.17. Isotherms at T_1, T_2 and T_3 are shown in Figs. 4.18(a), (b) and (c).

b. Vertical sectional diagram

For vertical sectional diagram an opened-up diagram (Fig. 4.11(c)) is useful. First draw a straight line along the vertical section (along ①–B in Fig. 4.19). Now let us draw a vertical sectional diagram along ①–B. Points at which the straight line ①–B intersects phase boundaries (①, ②, ③, ④, ⑤ and B) are characteristic points in the vertical section. Now let us refer to Figs. 4.14(a), (b) and (c).

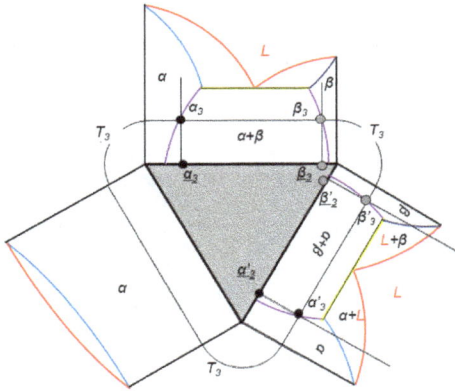

Fig. 4.17 Drawing a vertical section along ①–B.

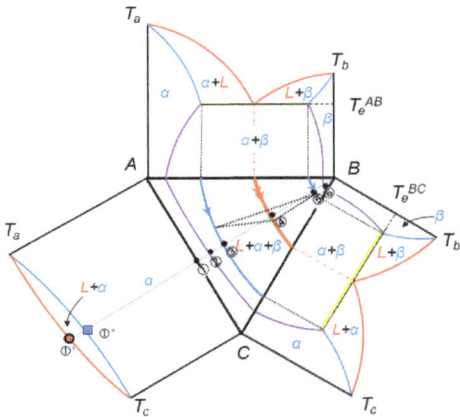

Fig. 4.19 Drawing a vertical section along ①–B.

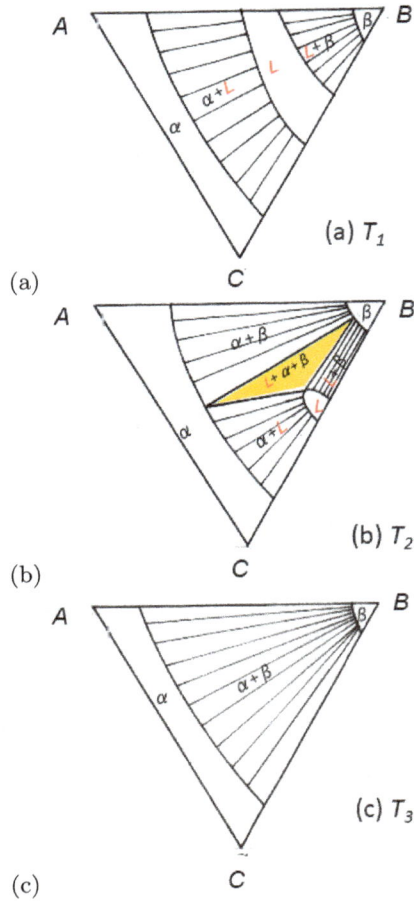

Fig. 4.18 Isotherms at (a) T_1, (b) T_2 and (c) T_3.

First, the upper top layer of PBS's is shown in Fig. 4.20(a). Here, out of ①–⑥ and B in Fig. 4.19, only ①, ④ and B are relevant, so that ① and ④ will be marked by **❶** and **❹**. Obviously from A-C binary system, **❶**'s are projections of two points ①′ and ①″ (temperature of ①′ > temperature of ①″: hereinafter denoted by ①′ > ①″), so that **❶** = ①′. By contrast, B corresponds to the melting point of B (B = T_b) (B is a pure metal). **❹** corresponds to L of the $(L + \alpha + \beta)$ three-phase triangle, that is, the point of intersection of two liquidus and top of the roofs (**❹** = L' (see Fig. 4.12)). Thus, the phase boundary obtained from Fig. 4.14(a) is indicated by a red line in Fig. 4.21 (drawn in red since it corresponds to liquidus).

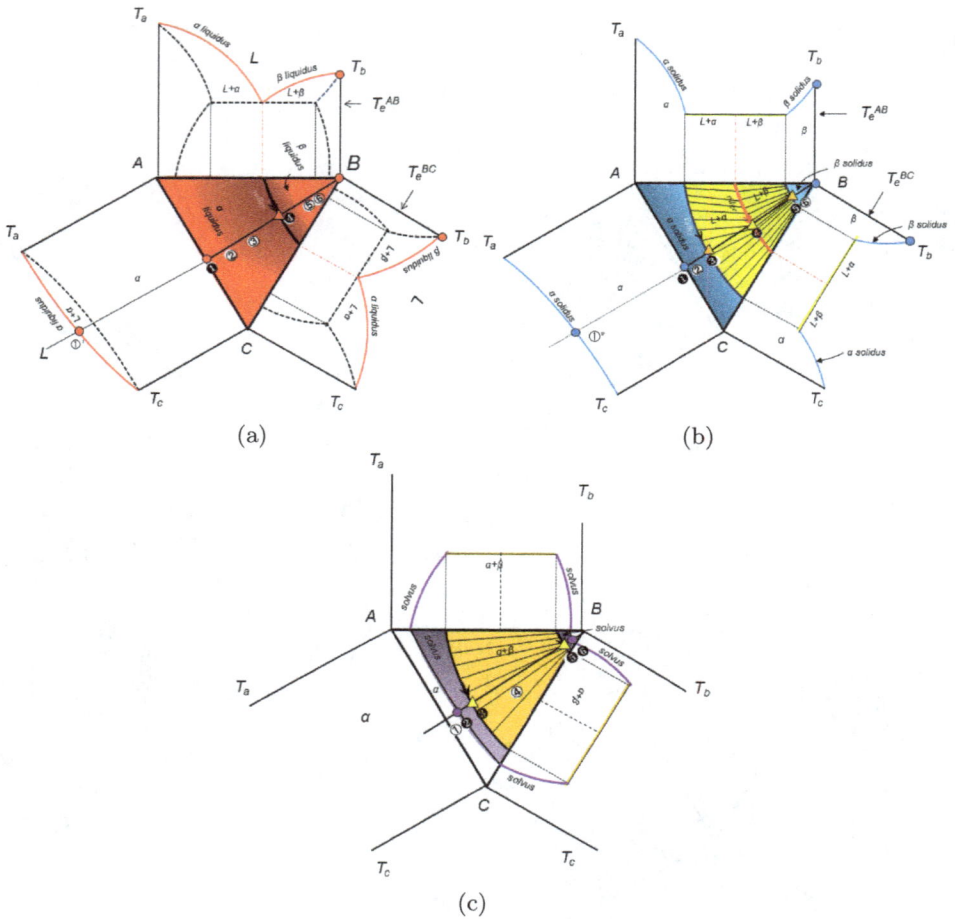

Fig. 4.20 Drawing a vertical section along ①–B. (a) Procedure ①, 1st PBS. (b) Procedure ②, 2nd PBS. (c) Procedure ③, 3rd PBS.

Next consider Fig. 4.20(b), where characteristic points are ❶, ❸, ❹, ❺ and B. Here, ❶ = ①″, B = T_b. Therefore, phase boundaries are as shown by blue lines in Fig. 4.21. Here ①″–❸ and ❺–T_b are solidus and drawn in navy blue. ❸–❹ and ❹–❺ correspond to roofs of the $(L + \alpha + \beta)$ three-phase triangle (Fig. 4.12), and drawn in yellow.

Lastly, from Fig. 4.20(c) characteristic points ❷, ❸, ❺ and ❻ are obvious. Connect these points in Fig. 4.21. Here, ❸–❺ corresponds to the lower broad face of a roof-shaped $(L + \alpha + \beta)$ three-phase triangle and drawn in ambre. ❷–❸ and ❺–❻ correspond to the solvus and drawn in purple. Filling the names of phases in single-phase fields, the vertical section along ❶–B is obtained as shown in Fig. 4.21.

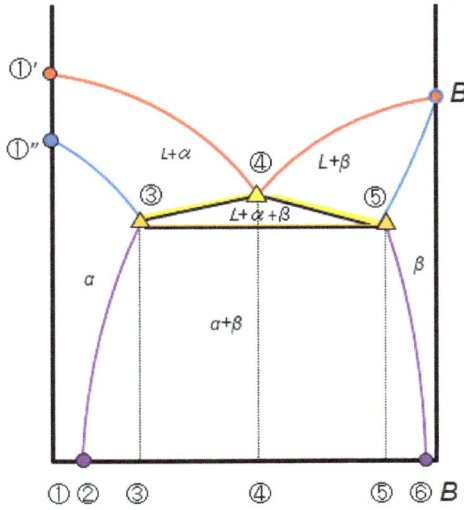

Fig. 4.21 Vertical section along ①–E (see Fig. 4.19).

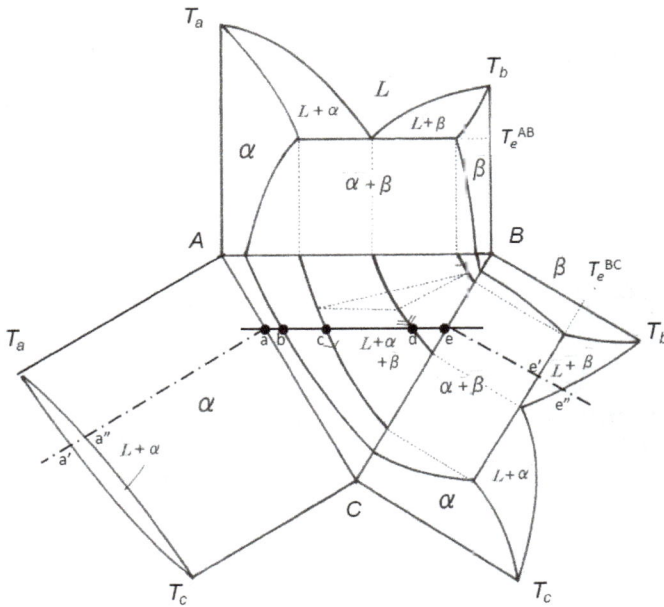

Fig. 4.22 Drawing a vertical section along **a-e**.

In the foregoing discussion, referring to the space diagrams shown in Figs. 4.13(a), (b) and (c) would be useful.

Exercise. Draw a vertical diagram along **a-e** in Fig. 4.22.

4.2.3   Simple peritectic system

A space diagram and an opened-up diagram of a ternary phase system of peritectic type are shown in Figs. 4.23(a), (b) and (c), respectively. Shaded fields in Figs. 4.23(a) and (b) correspond to peritectic reactions in the A-B and B-C binary systems, respectively. The key point of a ternary peritectic system is that three-phase triangle $(L+\alpha+\beta)$ is convex downward as shown in Fig. 4.24 (compare with $(L+\alpha+\beta)$ of a ternary system of a eutectic type (in Fig. 4.12)).

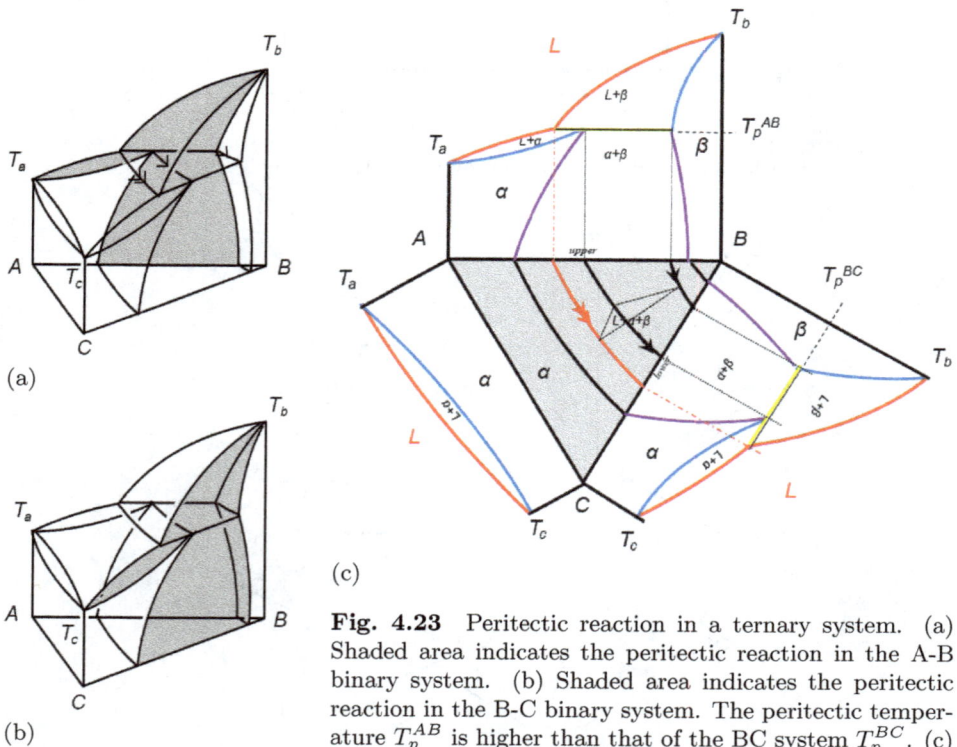

(a)

(b)

(c)

Fig. 4.23 Peritectic reaction in a ternary system. (a) Shaded area indicates the peritectic reaction in the A-B binary system. (b) Shaded area indicates the peritectic reaction in the B-C binary system. The peritectic temperature T_p^{AB} is higher than that of the BC system T_p^{BC}. (c) Projection of an opened-up space diagram.

Exercise. The roofs of the 3-phase triangles in the eutectic and peritectic systems are reversed. Discuss this reason.

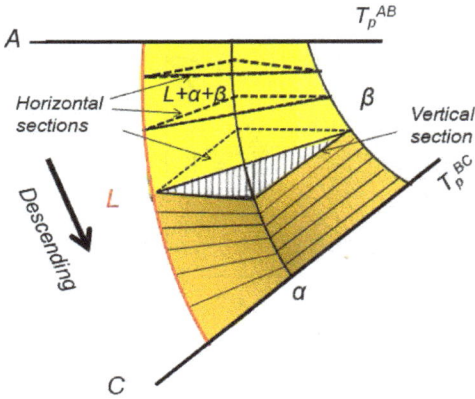

Fig. 4.24 Three-phase triangle ($L + \alpha + \beta$) in a peritectic system. T_p^{AB} is the peritectic temperature of the A-B binary peritectic system and T_p^{BC} is the peritectic temperature of the B-C binary peritectic system. T_p^{AB} is higher than T_p^{BC}.

Now let us decompose phase-boundary surface (PBS's) from high temperature layer by layer (Fig. 4.25, Fig. 4.26). The first PBS is the liquidus T_b-T_p^{AB}-T_p^{BC} and T_a-T_p^{AB}-T_p^{BC}-T_c. Here $T_p^{AB} > T_p^{BC}$, so that T_p^{AB}-T_p^{BC} which divides the liquidus surface into two (T_b-T_p^{AB}-T_p^{BC} and T_a-T_p^{AB}-T_p^{BC}-T_c) descends from AB side to BC side.

The next PBS's are solidus surface of B and the broad surface of ($L + \alpha + \beta$) three-phase triangle (convex downward). The third layer is a pair of narrow faces of ($L + \alpha + \beta$) three-phase triangle. The fourth layer is solidus surface of A-C binary side T_a-T_p^{AB}-T_p^{BC}-T_c. Lastly solvus surfaces of α and β phases appear.

The result is summarized in Table 4.3.

Table 4.3

	α phase side in the AC binary system	β-phase side		
Layer 1	Liquidus surface T_a-T_p^{AB}-T_p^{BC}-T_c	Liquidus surface T_b-T_p^{AB}-T_p^{BC}		
Layer 2		Broad face of ($L + \alpha + \beta$) (convex downward)		Solidus surface of β
Layer 3		One narrow face of ($L + \alpha + \beta$) (convex downward)	The other narrow face of ($L + \alpha + \beta$) (convex downward)	
Layer 4	Solidus surface of α			
Layer 5	Solvus surface of α			Solvus surface of β

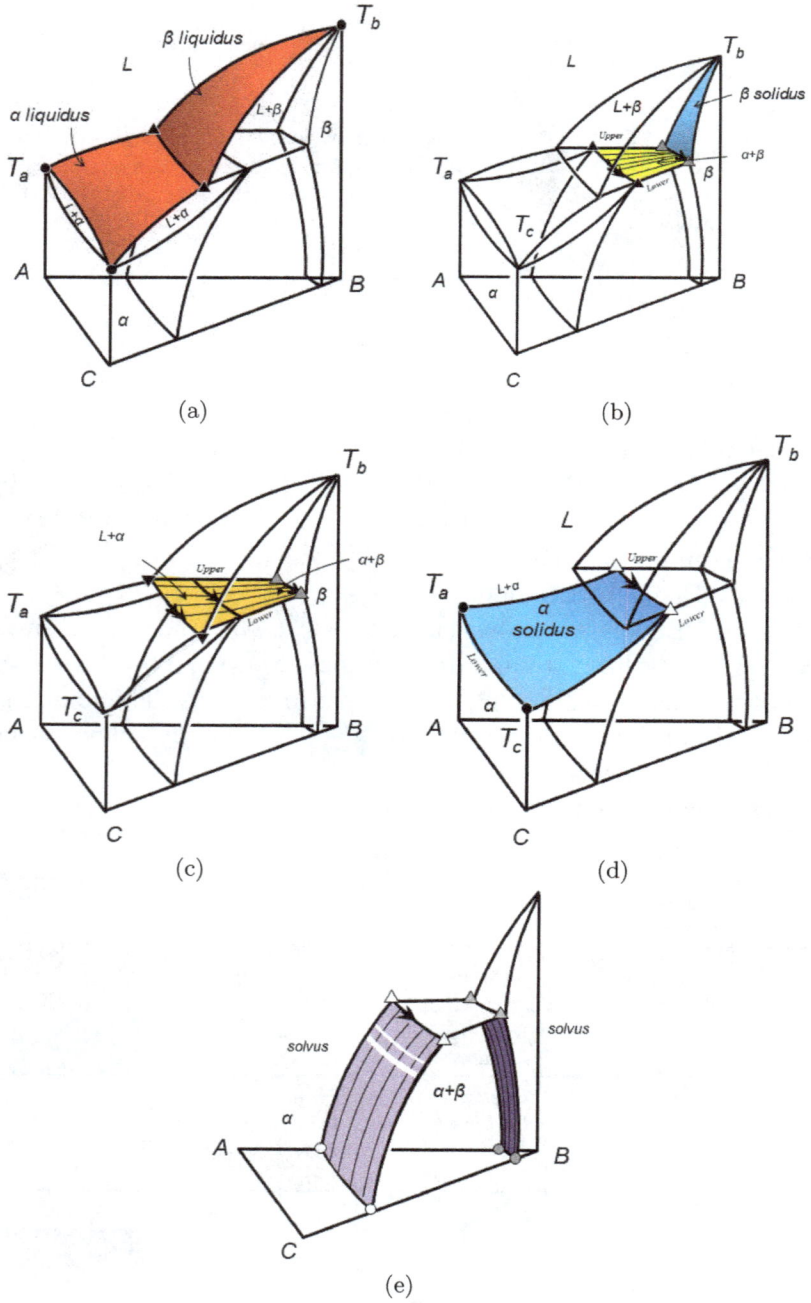

Fig. 4.25 (a) 1st PBS, (b) 2nd PBS, (c) 3rd PBS, (4) 4th PBS, (e) 5th PBS.

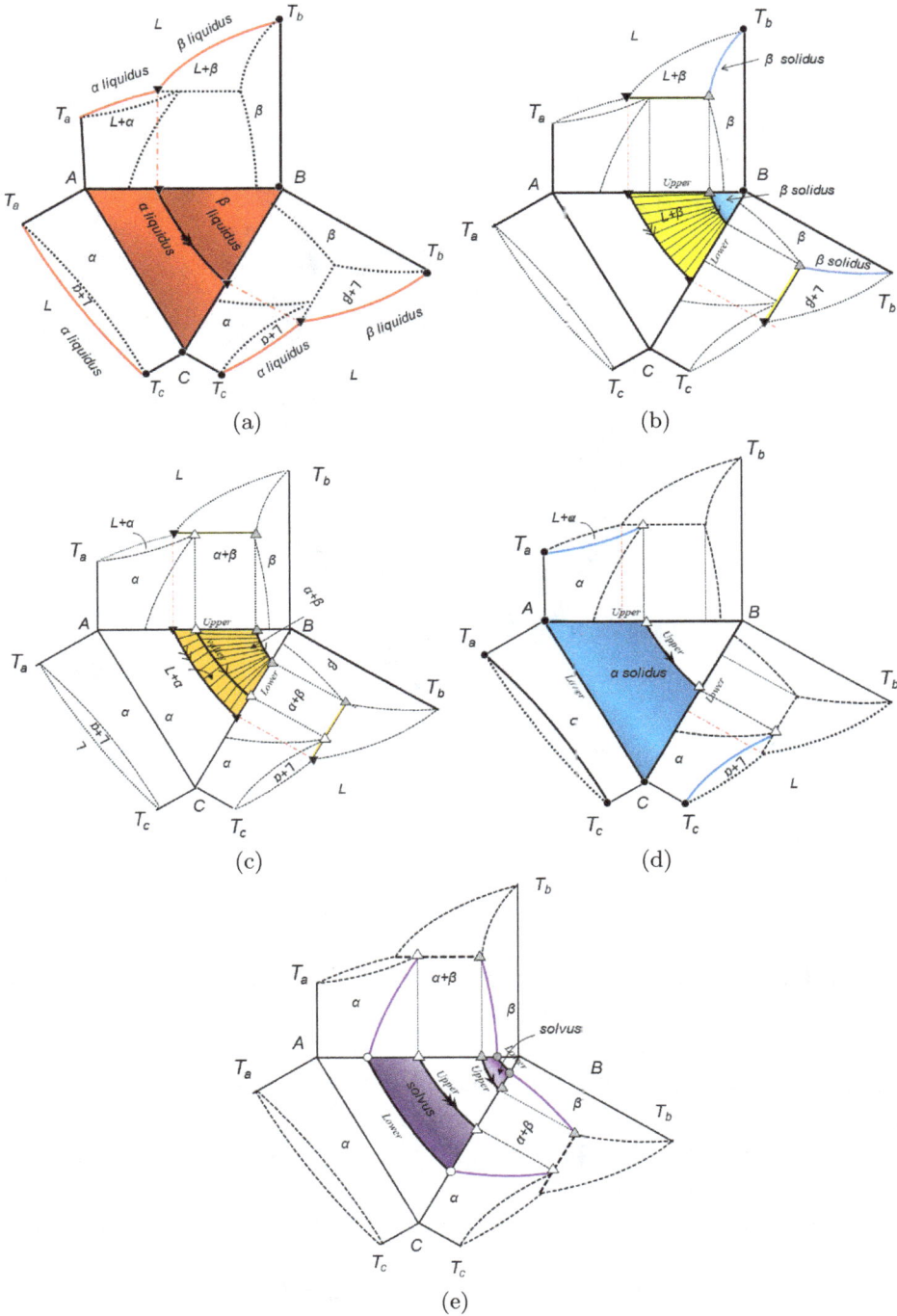

Fig. 4.26 (a) 1st PBS, (b) 2nd PBS, (c) 3rd PBS, (d) 4th PBS, (e) 5th PBS.

a. Isotherms

Procedures similar to those in the eutectic ternary system are applied to $T_1 \sim T_7$ in Fig. 4.27, then isotherms at the respective temperatures are obtained as shown in Figs. 4.28(a)–(g).

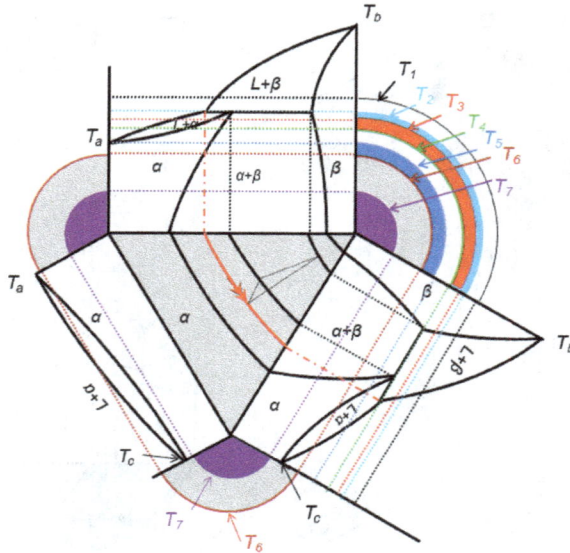

Fig. 4.27   Drawing isotherms.

b. Vertical sections

Now let us draw a vertical section along ①–B in Fig. 4.29.

Points where ①–B intersects with phase boundaries (①, ②, ③, ④, ⑤, ⑥ and B) are characteristic points in the vertical section. Now let refer to Fig. 4.25.

As is obvious from the binary A-C system, ❶ in the first layer in Fig. 4.30(a) are projections of intersections with liquidus ①′(●) and solidus ①″ (□). However, since the first layer is under consideration now, ❶ = ①′. Figure 4.26(a) shows that ❶ (①′)–❸ is the liquidus of α and that ❸–B is the liquidus of β and corresponds to the red line in Fig. 4.30(a). Next, in the second layer (Fig. 4.30(b)) ❸–❺ corresponds to the upper broad face of $(L + \alpha + \beta)$ three-phase triangle as can be seen from Fig. 4.25(b), and ❺–B is the solidus of β. Figure 4.25(c) shows that ❸–❹ and ❹–❺ are two lower, narrow faces of $(L + \alpha + \beta)$ three-phase triangle, flanking $(L + \alpha)$ and $(\alpha + \beta)$, respectively. In the third layer (Fig. 4.30(d)), ❶ = ①″ (= solidus

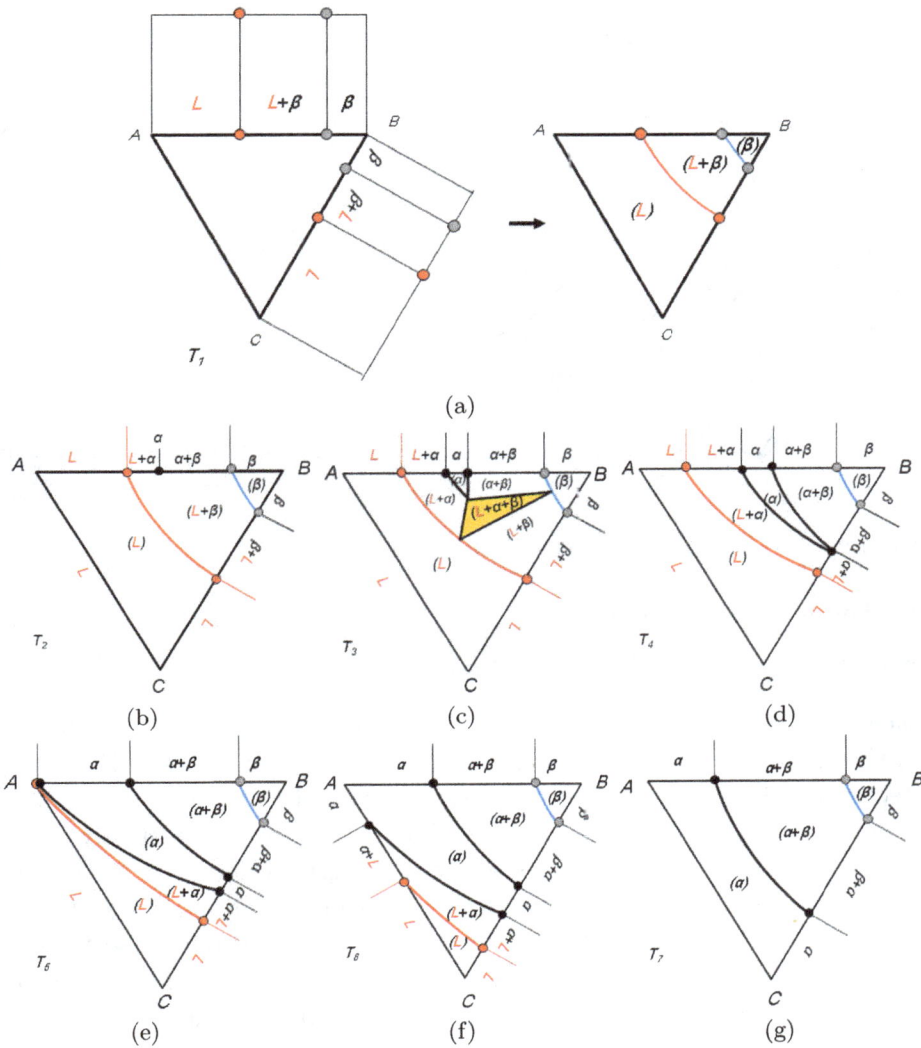

Fig. 4.28 Isotherms at (a) temperature T_1, (b) temperature T_2, (c) temperature T_3, (d) temperature T_4, (e) temperature T_5, (f) temperature T_6, (g) temperature T_7.

of α in the AC binary system, $\square$), so that ❶ (①″ = $\square$)-❹ is the solidus surface of α. Finally, in the fourth layer (Fig. 4.30(e)) ❷-❸ and ❺-❻ are solvus lines of α and β, respectively (refer to Fig. 4.25(e)).

The resultant vertical section is shown in Fig. 4.31.

Fig. 4.31   A vertical section along ①–B (Fig. 4.29).

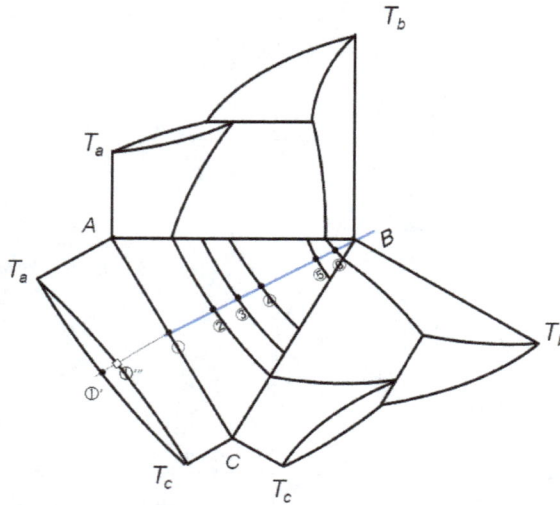

Fig. 4.29   Drawing a vertical section along ①–B.

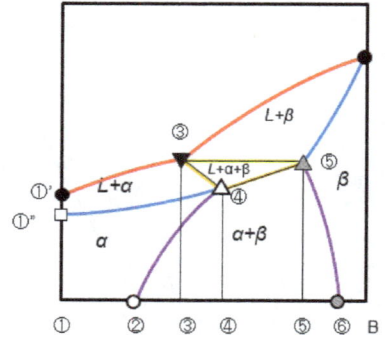

4.3   Complex Ternary Systems

4.3.1   Ternary eutectic system (Class I)

Space diagrams of a ternary eutectic system are shown in Figs. 4.32(a), (b) and the opened-up one in Fig. 4.32(c). The shaded field in Fig. 4.32(a) is the eutectic reaction in the A-B binary system. Similarly, those in Fig. 4.32(b) are eutectic reactions in B-C and C-A binary systems, respectively. All of the A-B, B-C and C-A binary systems are eutectic systems and each of the eutectic temperatures descends towards the inside of the ternary system accompanying $(L + \alpha + \beta)$, $(L + \beta + \gamma)$, $(L + \gamma + \alpha)$ three-phase triangles (tie-angles) (Fig. 4.32(c)). These bottoms of the three-phase triangles join on a ternary eutectic plane at a certain temperature (ternary eutectic temperature) to form the four-phase eutectic field ((α-β-γ) triangle $+L$ at the center). At this temperature four phases, i.e., $L + \alpha + \beta + \gamma$ are coexistent, so that $f = 0$. That is, the invariant reaction occurs. Once L disappears, the remaining three phases, i.e., $(\alpha + \beta + \gamma)$ are coexistent, resulting in $f = 1$ again and temperature lowers.

Phase boundary surfaces are shown in Figs. 4.33(a)–(d) and Figs. 4.34(a)–(d).

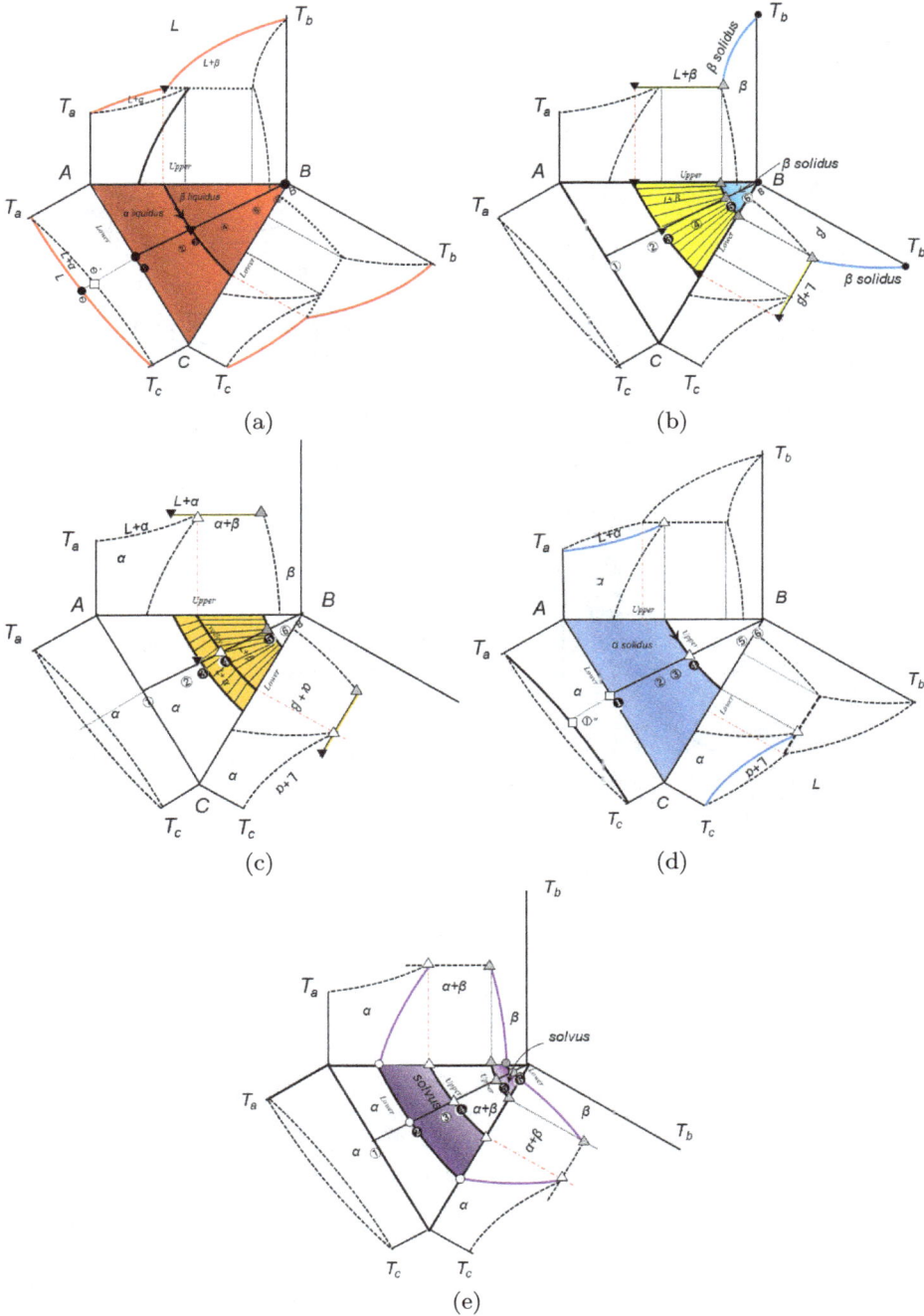

Fig. 4.30 (a) 1st PBS, (b) 2nd PBS, (c) 3rd PBS, (d) 4th PBS, (e) 5th PBS.

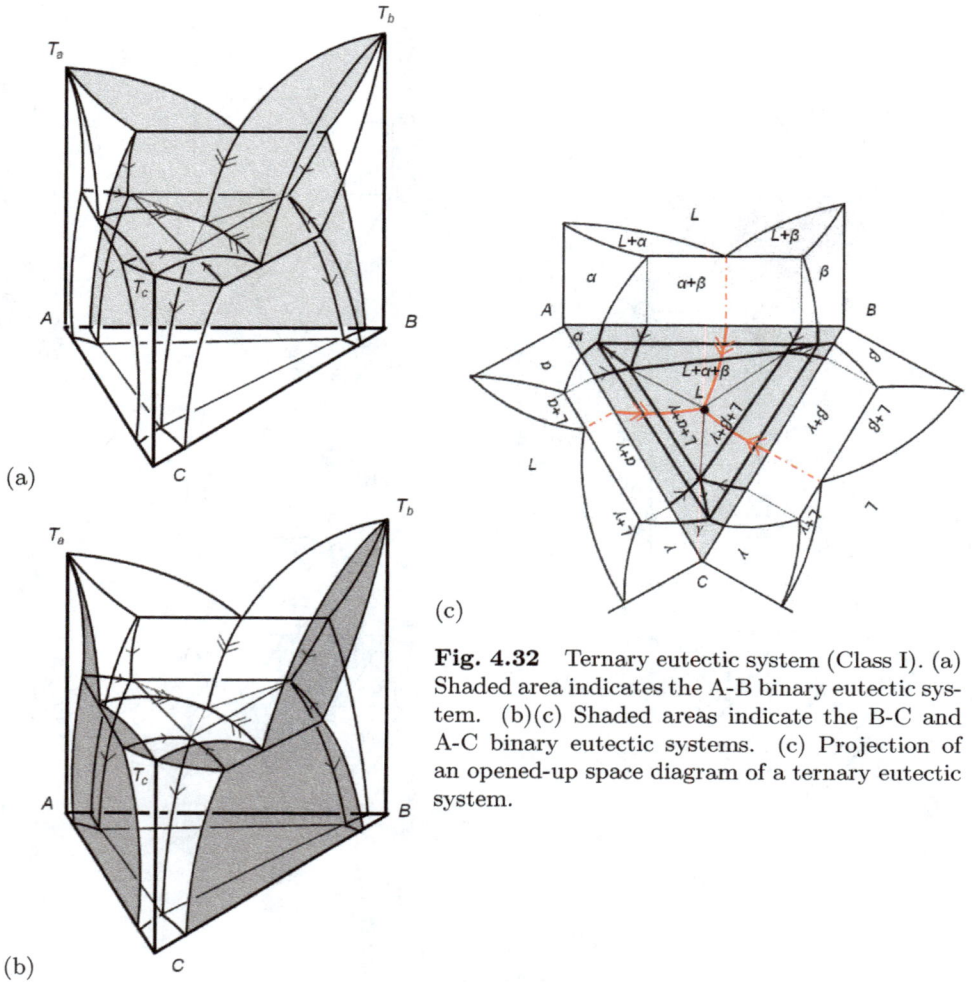

(a)

(b)

(c)

Fig. 4.32 Ternary eutectic system (Class I). (a) Shaded area indicates the A-B binary eutectic system. (b)(c) Shaded areas indicate the B-C and A-C binary eutectic systems. (c) Projection of an opened-up space diagram of a ternary eutectic system.

a. Isotherms

Let the eutectic temperatures in the A-B, B-C and C-A binary systems be T_e^{AB}, T_e^{BC} and T_e^{CA}, respectively and let the ternary eutectic temperature be T_e^{III}, where $T_e^{AB} > T_e^{BC} > T_e^{CA} > T_e^{III}$. Isotherms at $T_1 \sim T_6$ in Fig. 4.35 are shown in Figs. 4.36(a)–(f).

At $T_1 = T_e^{AB}$ (Fig. 4.36(a)), the eutectic reaction appears in the A-B binary system, but not yet in the B-C nor C-A binary systems. At $T_2 = T_e^{BC}$ (Fig. 4.36(b)) the eutectic reaction appears in the B-C binary system, so does the eutectic reaction at $T_3 = T_e^{CA}$ (Fig. 4.36(c)). These three eutectic

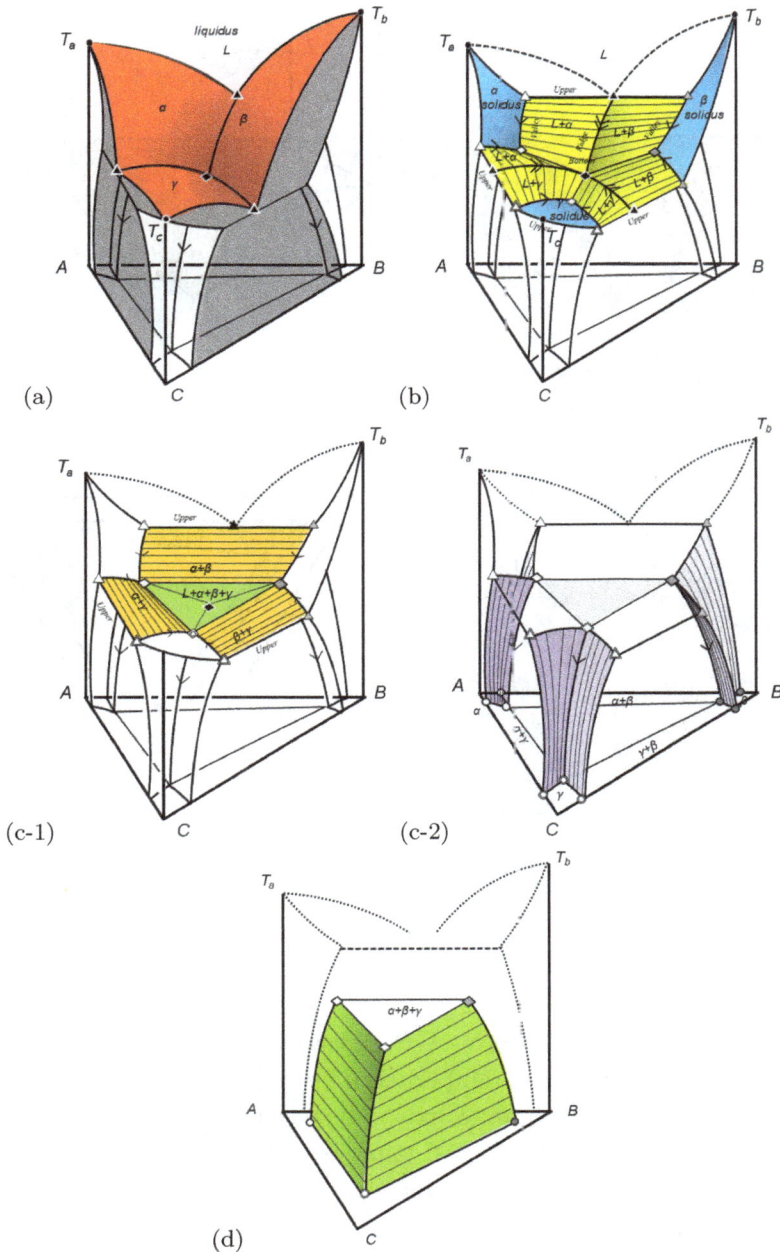

Fig. 4.33 Ternary eutectic system (Class I). (a) 1st PBS, (b) 2nd PBS, (c-1) 3rd PBS, (c-2) 4th PBS, (d) 5th PBS.

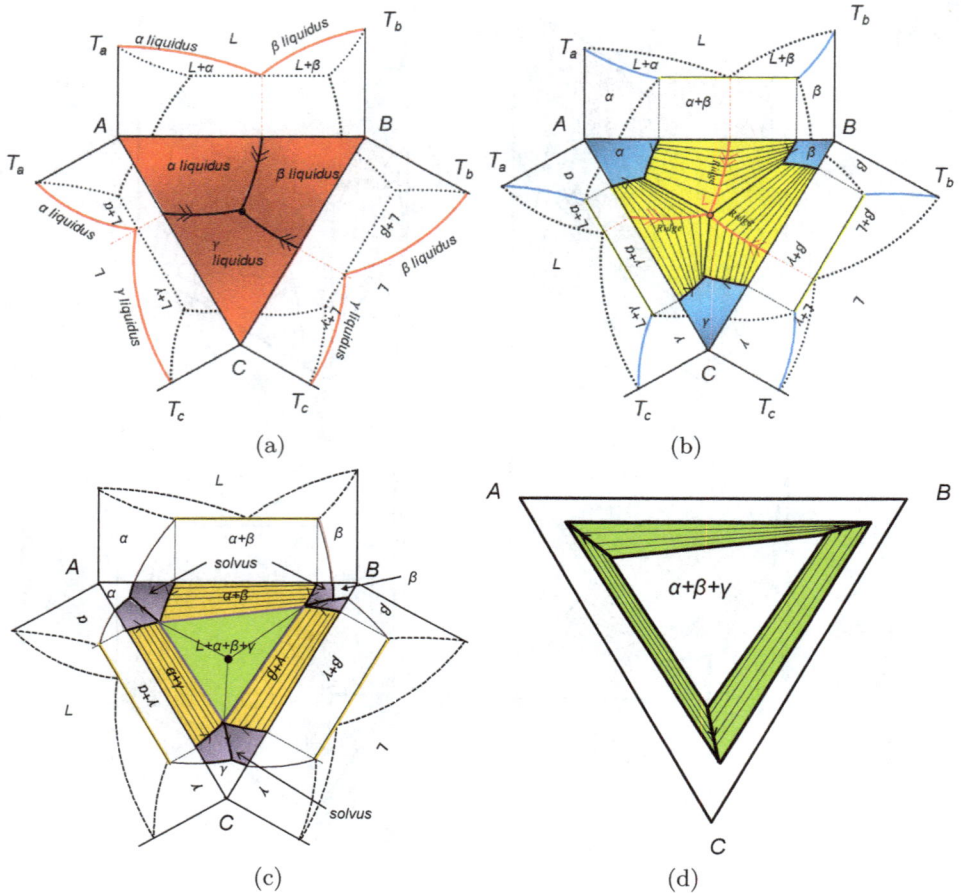

Fig. 4.34 Ternary eutectic system (Class I). (a) 1st PBS, (b) 2nd PBS, (c) 3rd PBS, (d) 4th PBS. Here, (c-1) and (c-2) in Fig. 4.33 are combined together in (c).

reactions go downward inside of A-B-C ternary system (Figs. 4.36(d)(e)), reaching the ternary eutectic temperature $T_5 = T_e^{III}$ (Fig. 4.36(e)). At T_e^{III} four phases, i.e., $L+\alpha+\beta+\gamma$ are coexistent, so that the reaction $L \to \alpha+\beta+\gamma$ is an invariant reaction with $f = 0$ (that is, the temperature during which this reaction takes place remains unchanged). When the reaction finishes, that is, L disappears, three phases, i.e., $\alpha+\beta+\gamma$ remain (Fig. 4.36(f)), with $f = 1$ again.

b. Vertical sections

Let us draw a vertical section along ①–⑨ in Fig. 4.37.

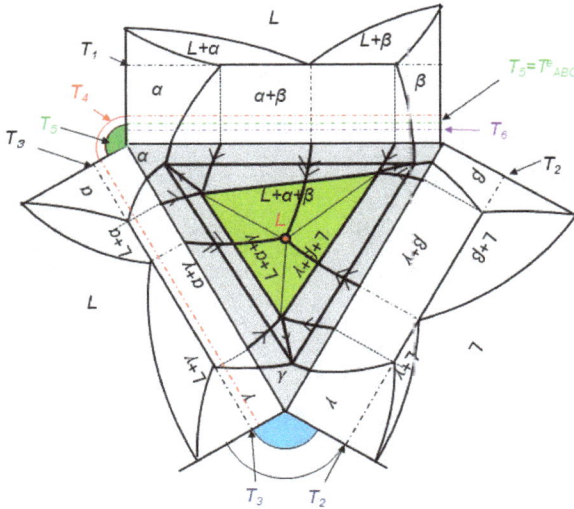

Fig. 4.35 Drawing isotherms (Class I).

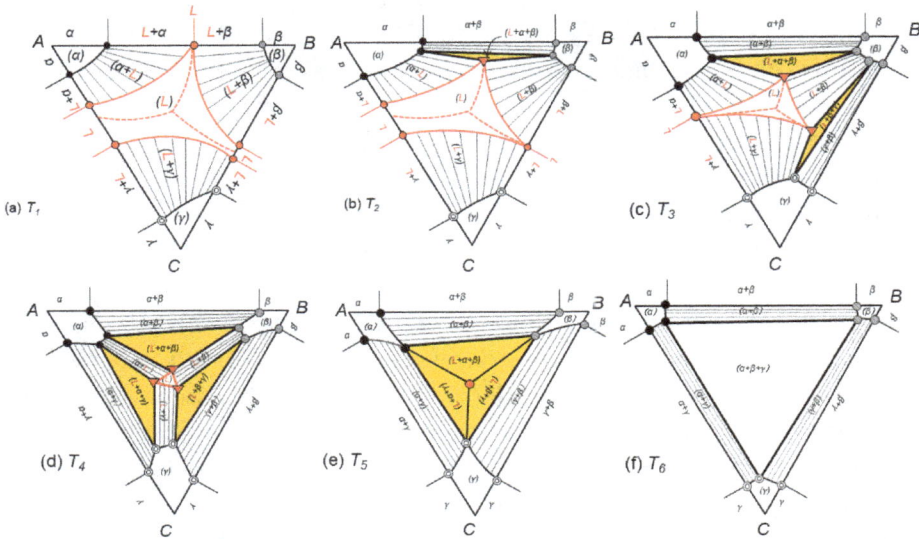

Fig. 4.36 Ternary eutectic system (Class I). Isotherms at (a) T_1, (b) T_2, (c) T_3, (d) T_4, (e) T_5, (f) T_6.

Characteristic points in the first surface (Fig. 4.38(a)) are ❶, ❺ and ❾, among which obviously ❶ = ①′ (liquidus). Similarly ❾ = ⑨′. ❶—❺ is liquidus surface of α, ❺—❾ is the liquidus surface of β.

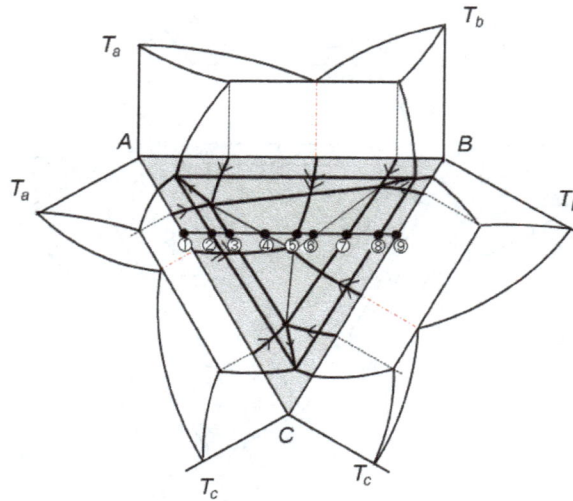

Fig. 4.37 Vertical sections (Class I).

On the second surface (Fig. 4.38(b)) **❶**, **❹**, **❺**, **❻** and **❾** are characteristic points, where obviously **❶** = ①″ and **❾** = ⑨″. **❶**–**❹** is the roof $(L+\alpha)$ of the three-phase triangle $(L+\alpha+\gamma)$, so that **❶** ◺ **❹**. **❹**–**❺** is the roof $(L+\alpha)$ of the three-phase triangle $(L+\alpha+\beta)$. **❺**–**❻** is the roof $(L+\beta)$ of $(L+\alpha+\beta)$. Therefore, **❹** ◿ **❺** ◺ **❻** with **❺** at the summit (indicated by ▲). **❻**–**❾** is the roof $(L+\beta)$ of $(L+\beta+\gamma)$ and **❻** ◿ **❾**.

On the fourth surface (Fig. 4.38(c)) **❶**, **❸**, **❼** and **❾** are characteristic points where **❶** = ①″ and **❾** = ⑨″. **❶**–**❸** is the broad face $(\alpha+\gamma)$ of the three-phase triangle $(L+\alpha+\gamma)$, and **❼**–**❾** is the broad face $(\beta+\gamma)$ of the three-phase triangle $(L+\beta+\gamma)$. **❸**–**❼** corresponds to the invariant reaction of the ternary eutectic reaction $(L+\alpha+\beta+\gamma)$ and is to be expressed by a horizontal line. Therefore, **❶** ◺ **❸** → **❼** ◿ **❾**. Among them, **❸**–④ corresponds to the three-phase triangle $(L+\alpha+\gamma)$. Similarly, ④–⑥ corresponds to the three-phase triangle $(L+\alpha+\beta)$ and ⑥–**❼** corresponds to the broad face of the three-phase triangle $(L+\gamma+\beta)$.

On the fifth surface (Fig. 4.38(d)) **❷**, **❸**, **❼** and **❽** are characteristic points. **❸**–**❷**, **❼**–**❽** are the walls of a pyramid of the three-phase $(\alpha+\beta+\gamma)$, with **❸** ◺ **❷** and **❼** ◺ **❽**.

The resultant vertical section along ①–⑨ is shown in Fig. 4.39.

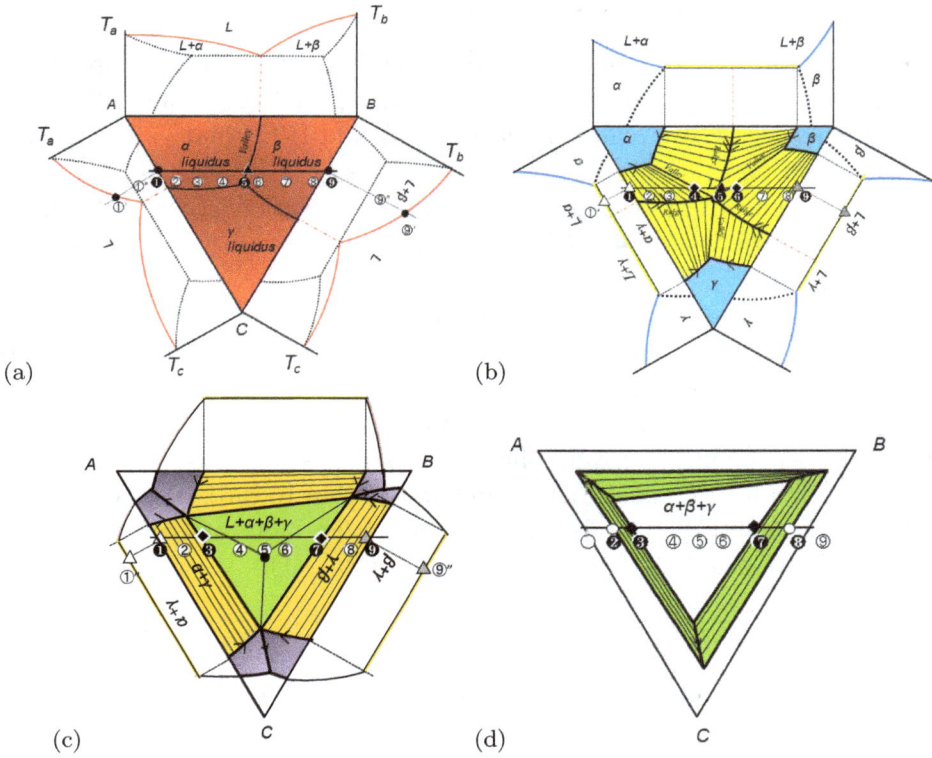

Fig. 4.38 Ternary eutectic system (Class I). (a) 1st PBS, (b) 2nd PBS, (c) 3rd PBS, (d) 4th PBS.

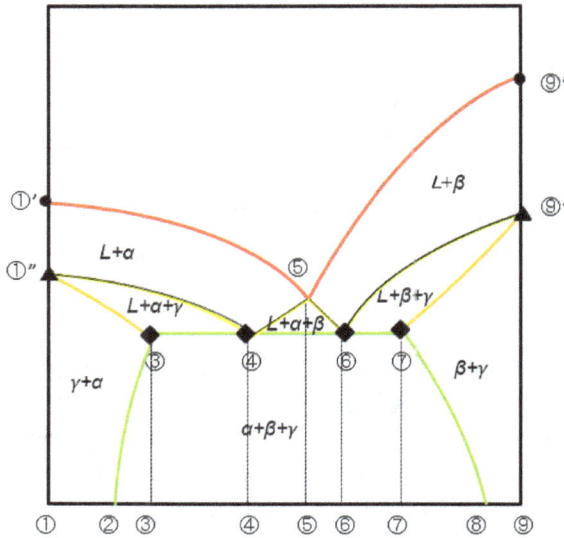

Fig. 4.39 Vertical section along ①–⑨ in Fig. 4.37 (Class I).

4.3.2 Peritecto-eutectic system (Class II)

Space diagrams of a ternary peritecto-eutectic system (Class II) are shown in Figs. 4.40(a)(b) and an opened-up one in Fig. 4.40(c). The A-B binary system has a peritectic reaction and the C-A and B-C binary systems have eutectic reactions.

Shaded field in Fig. 4.40(a) is the peritectic reaction in the A-B binary system and those in Fig. 4.40(b) are eutectic reactions in the B-C and C-A binary systems.

The three-phase triangles $(L+\alpha+\beta)$ and $(L+\alpha+\gamma)$ descend towards the B-C binary system joining at a certain temperature (the ternary peritecto-eutectic temperature, T_{pe}^{III}) and form the four-phase invariant plane $(L+\alpha+\beta+\gamma)$. The point here is that liquid phase L is outside of the three-phase triangle $(\alpha+\beta+\gamma)$. This liquid phase L constitutes the three-phase triangle $(L+\beta+\gamma)$, which descends as a eutectic reaction towards the B-C binary system (Fig. 4.40(c)). The PBS's are shown from high temperature to low temperature in Figs. 4.41(a)–(f) and Figs. 4.42(a)–(f).

a. Isotherms

Isotherms at each of the temperatures shown in Fig. 4.43 are shown in Figs. 4.44(a)–(d).

b. Vertical sections

Let us draw a vertical section along ①–⑨ in Fig. 4.45.

Characteristic points in the first surface (Fig. 4.46(a)) are ❶, ❼ and ❾, where ❶ = ①′, ❾ = ⑨′. Inspection of the geometry of the liquidus surfaces shows that ❶ [↘] ❼, ❼ [↘] ❾.

On the second surface (Fig. 4.46(b)) ❶, ❹ and ❼ are characteristic points, where ❶ = ①″, and ❶–❹ is the roof $(L+\gamma)$ of the three-phase triangle (eutectic) $(L+\beta+\gamma)$, and ❹–❼ is a broader face $(L+\alpha)$ of the three-phase triangle (peritectic) $(L+\alpha+\beta)$. In other words, ❼ is the liquidus line of the three-phase triangle (peritectic) $(L+\alpha+\beta)$, so that ❶ [↘] ❹ [↗] ❼.

On the third surface (Fig. 4.46(c)) ❶, ❸, ❻ and ❼ are characteristic points, where ❶ = ①″ and ❶–❸ is the broader face $(\alpha+\gamma)$ of the three-phase triangle (eutectic) $(L+\alpha+\gamma)$. On the other hand, ❻–❼ is the narrow face $(L+\beta)$ underneath the three-phase triangle (peritectic) $(L+\alpha+\beta)$. ❸–❻ corresponds to the invariant reaction of the four-phase equilibrium $(L+\alpha+\beta+\gamma)$, and should be horizontal. Therefore, ❶ [↘] ❸–❻ [↗] ❼.

(a)

(b)

(c)

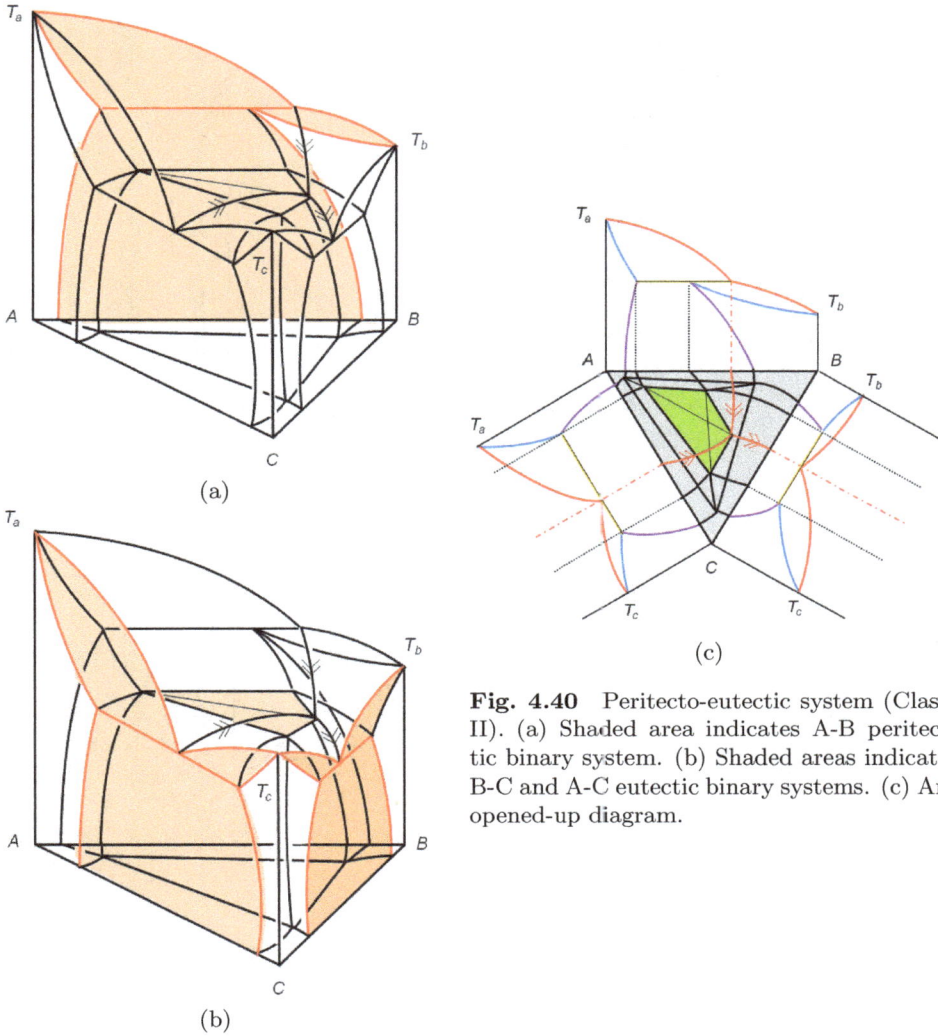

Fig. 4.40 Peritecto-eutectic system (Class II). (a) Shaded area indicates A-B peritectic binary system. (b) Shaded areas indicate B-C and A-C eutectic binary systems. (c) An opened-up diagram.

On the fourth surface (Fig. 4.46(d)) ❻ and ❾ are characteristic points, where ❾ = ⑨″, and ❻–❾ is the roof $(L + \beta)$ of the three-phase triangle (eutectic) $(L + \beta + \gamma)$. Therefore, ❻ ◨ ❾.

On the fifth surface (Fig. 4.46(e)) ❺ and ❾ are characteristic points, where ❺–❾ is the broader face $(\beta + \gamma)$ underneath the three-phase triangle (eutectic) $(L + \beta + \gamma)$. Therefore, ❺ ◨ ❾.

On the sixth surface (Fig. 4.46(f)) ❷, ❸, ❺ and ❽ are characteristic points, where ❷–❸, and ❺–❽ are walls of the three-phase triangle pyramid $(\alpha + \beta + \gamma)$. Therefore, ❷ ◨ ❸ and ❺ ◨ ❽.

The resultant vertical section is shown in Fig. 4.47.

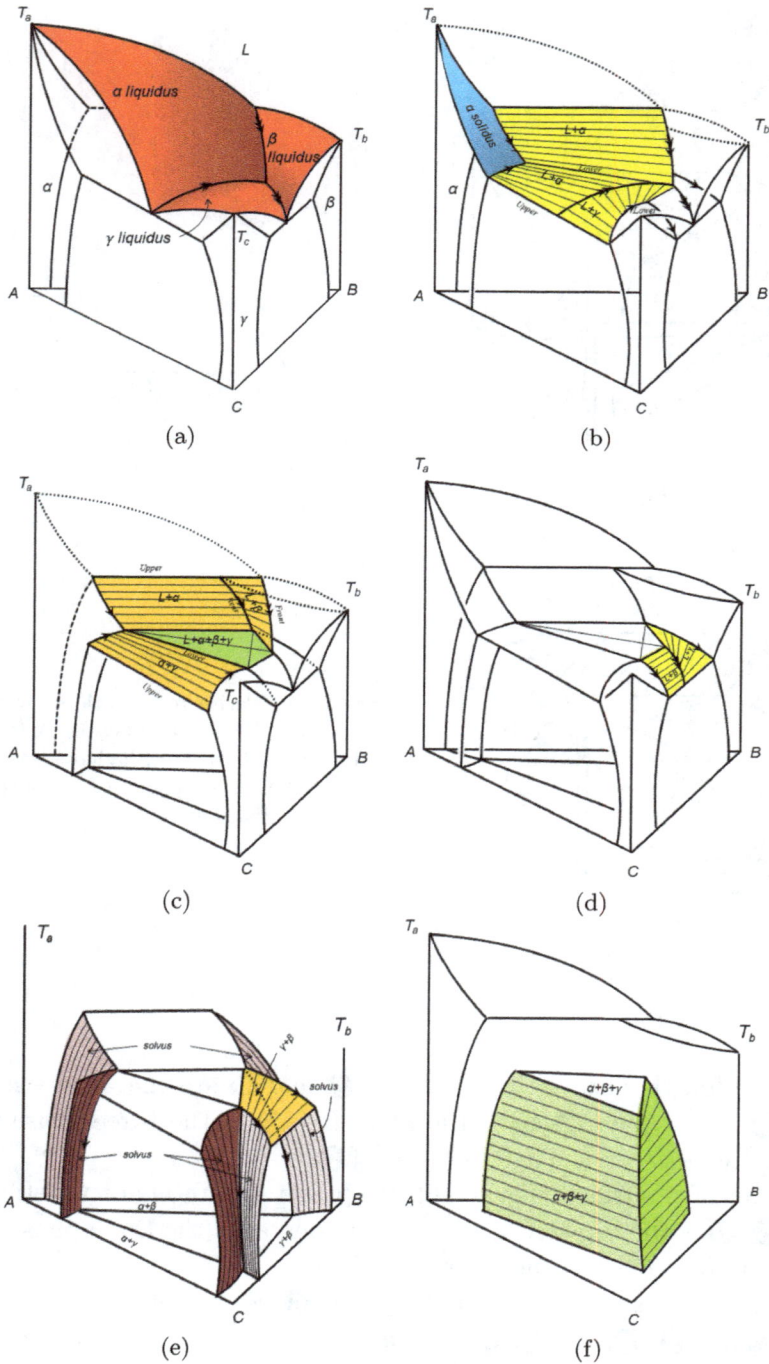

Fig. 4.41 Peritecto-eutectic system (Class II). (a) 1st PBS, (b) 2nd PBS, (c) 3rd PBS, (d) 4th PBS, (e) 5th PBS, (f) 6th PBS.

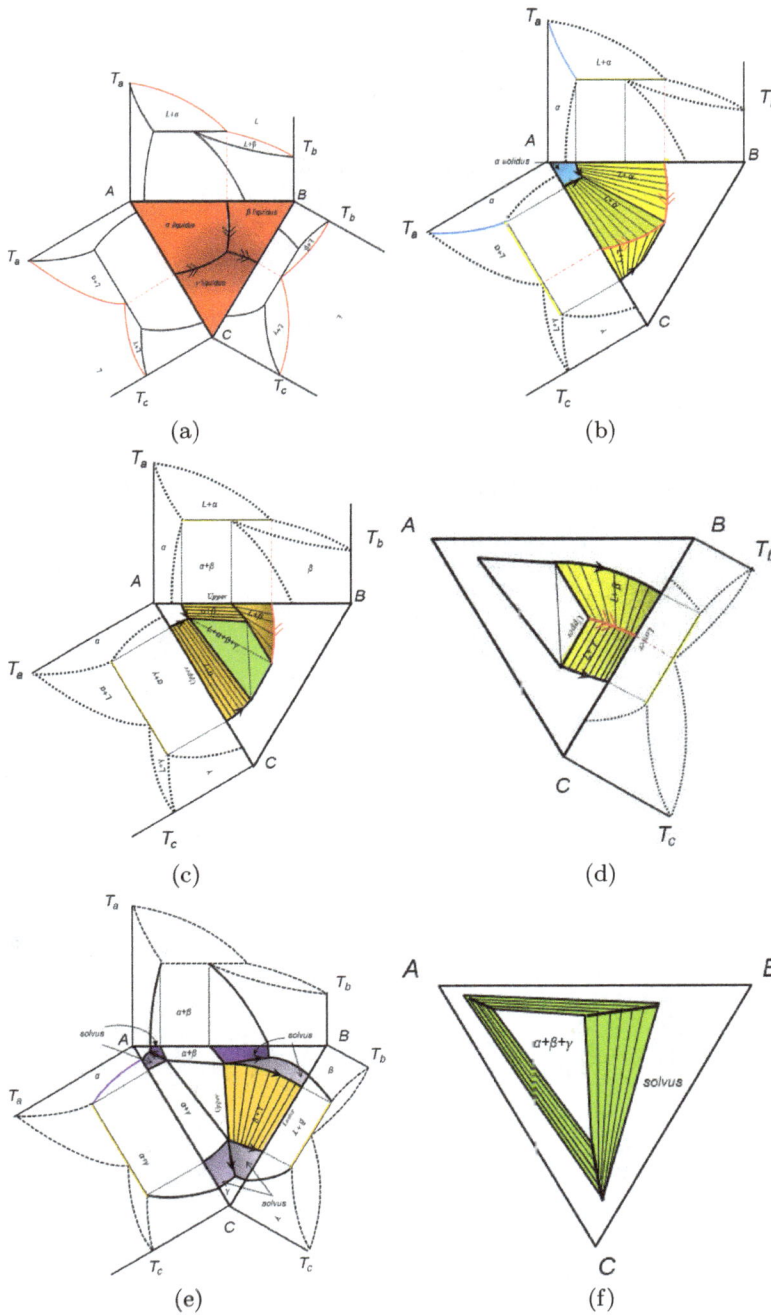

(a)

(b)

(c)

(d)

(e)

(f)

Fig. 4.42 Peritecto-eutectic system (Class II). (a) 1st PBS, (b) 2nd PBS, (c) 3rd PBS, (d) 4th PBS, (e) 5th PBS, (f) 6th PBS.

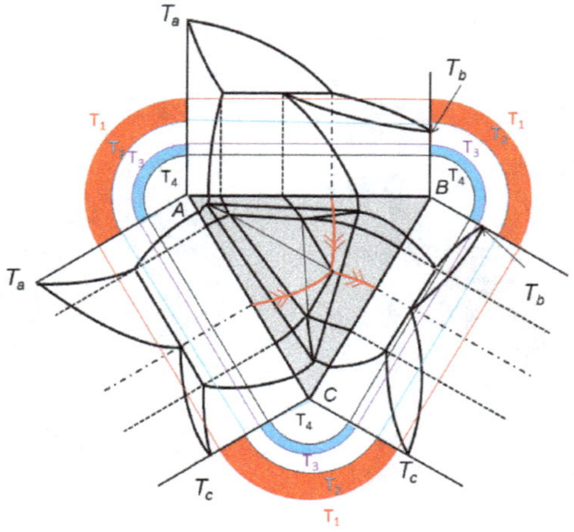

Fig. 4.43 Drawing isotherms (peritecto-eutectic system (Class II)).

(a)

(b)

(c)

(d)

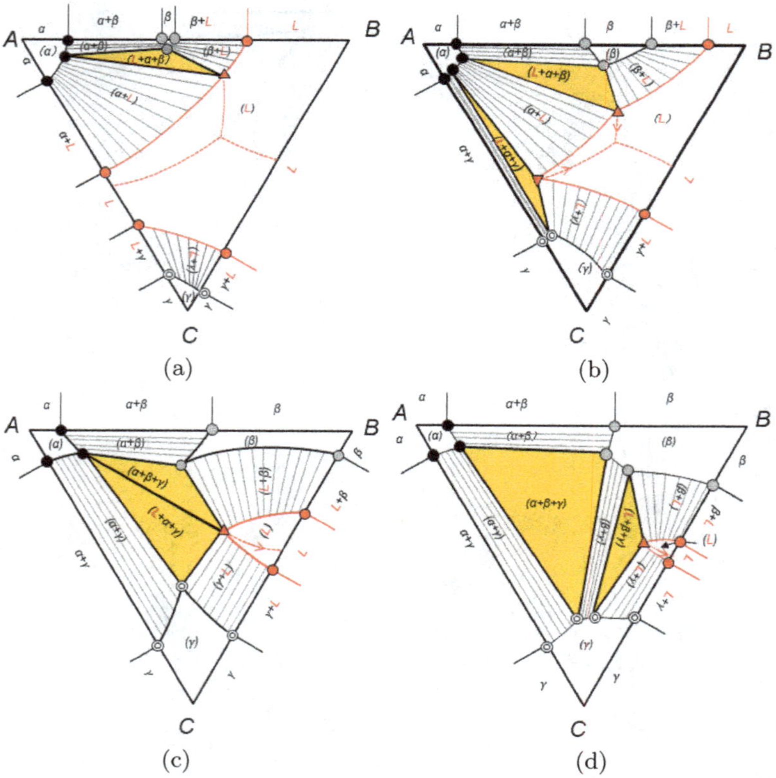

Fig. 4.44 Peritecto-eutectic system (Class II). Isotherms at (a) T_1, (b) T_2, (c) T_3, (d) T_4.

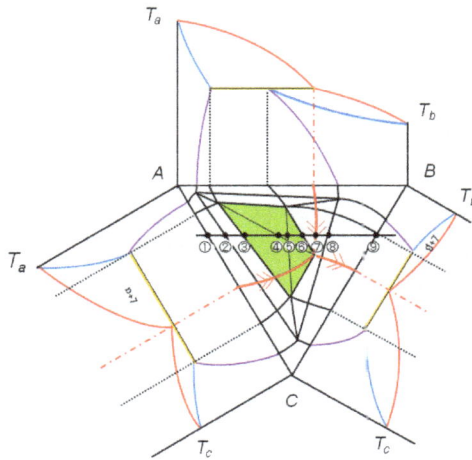

Fig. 4.45 Drawing a vertical section along ①–⑨ (Class II).

Exercise. The vertical sections along ② and ③ in Fig. 4.48(a) are shown in Figs. 4.48(b)(c). Confirm this. In (b) α-phase and in (c) β-phase are already located. Name the phases in the other fields.

4.3.3 Peritecto-eutectic system (Class III)

Figures 4.49(a), (b) and (c) show space diagrams and an opened-up figure of a peritecto-eutctic diagram (Class III). The A-B binary system has a eutectic reaction (shaded in Fig. 4.49(a)), and both of B-C and C-A binary systems have peritectic reactions (shaded in Fig. 4.49(b)). The eutectic reaction $(L + \alpha + \beta)$ in the A-B binary system descends toward the inside of the ternary system, and at the peritecto-eutectic temperature T, γ-phase appears inside the three-phase triangle $(L + \alpha + \beta)$. Then, γ and α-phases form a peritectic reaction with β-phase and descend towards the B-C and C-A binary systems (Fig. 4.49(c)).

Phase boundary surfaces are shown in Figs. 4.50(a)–(g) and Figs. 4.51(a)–(g) from high temperature layer by layer.

a. Isotherms

Figures 4.53(a)–(d) show the isotherms at each of the temperatures shown in Fig. 4.52.

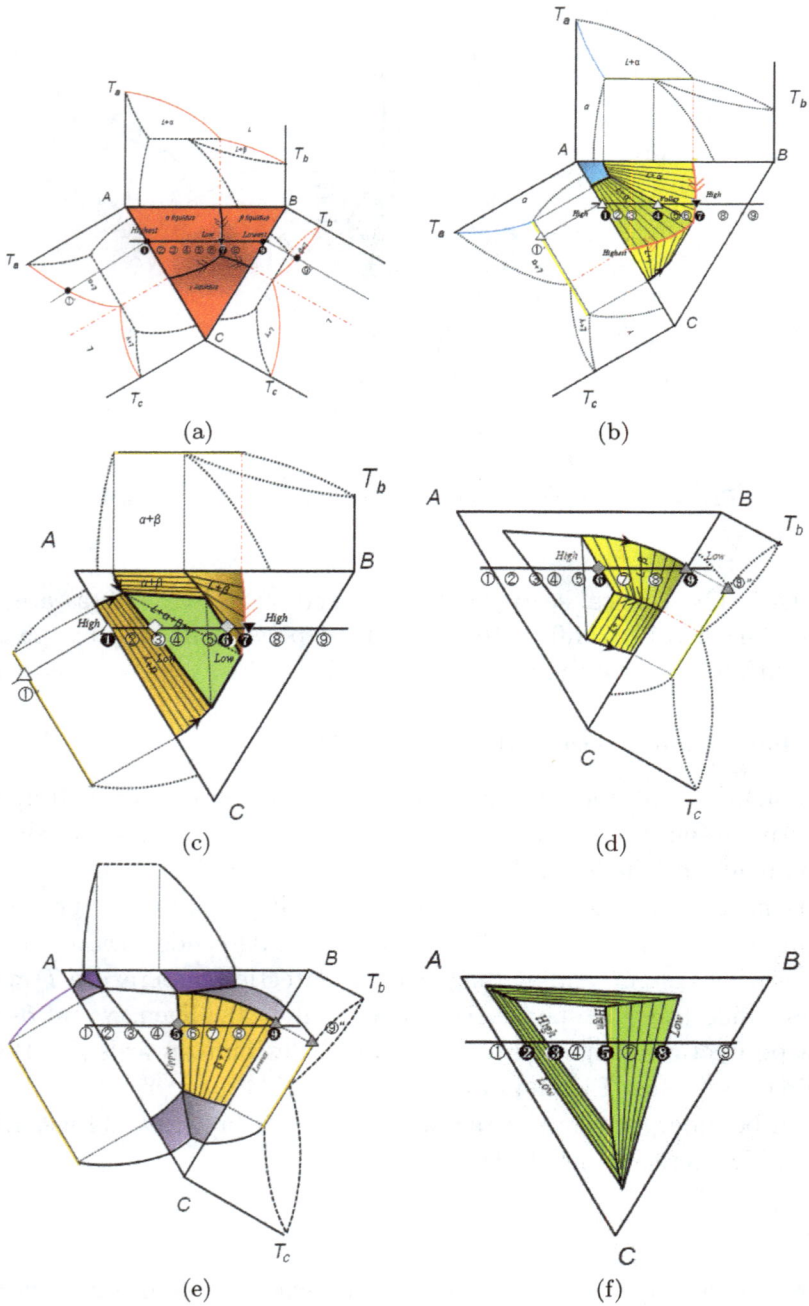

Fig. 4.46 Peritecto-eutectic system (Class II). (a) 1st PBS, (b) 2nd PBS, (c) 3rd PBS, (d) 4th PBS, (e) 5th PBS, (f) 6th PBS.

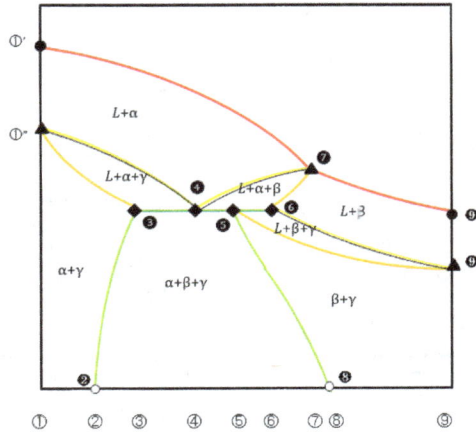

Fig. 4.47 Peritecto-eutectic system (Class II). Vertical section along ①–⑨ in Fig. 4.45.

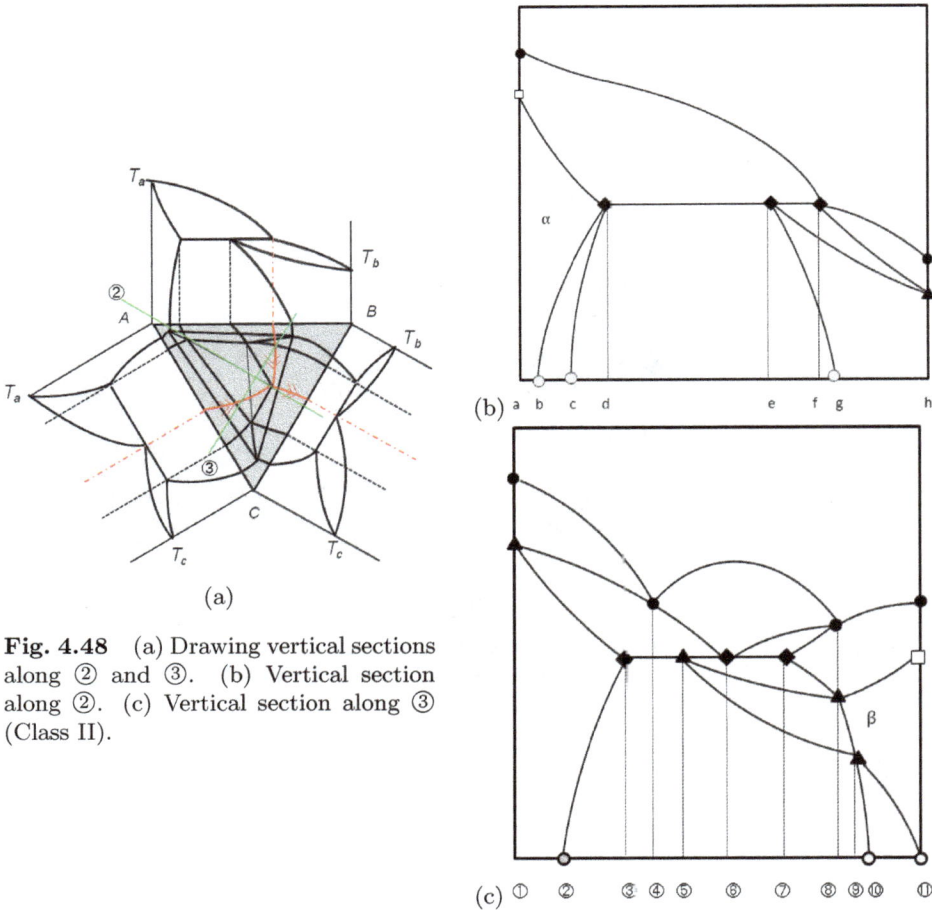

(a)

Fig. 4.48 (a) Drawing vertical sections along ② and ③. (b) Vertical section along ②. (c) Vertical section along ③ (Class II).

(a)   (b)

(c)

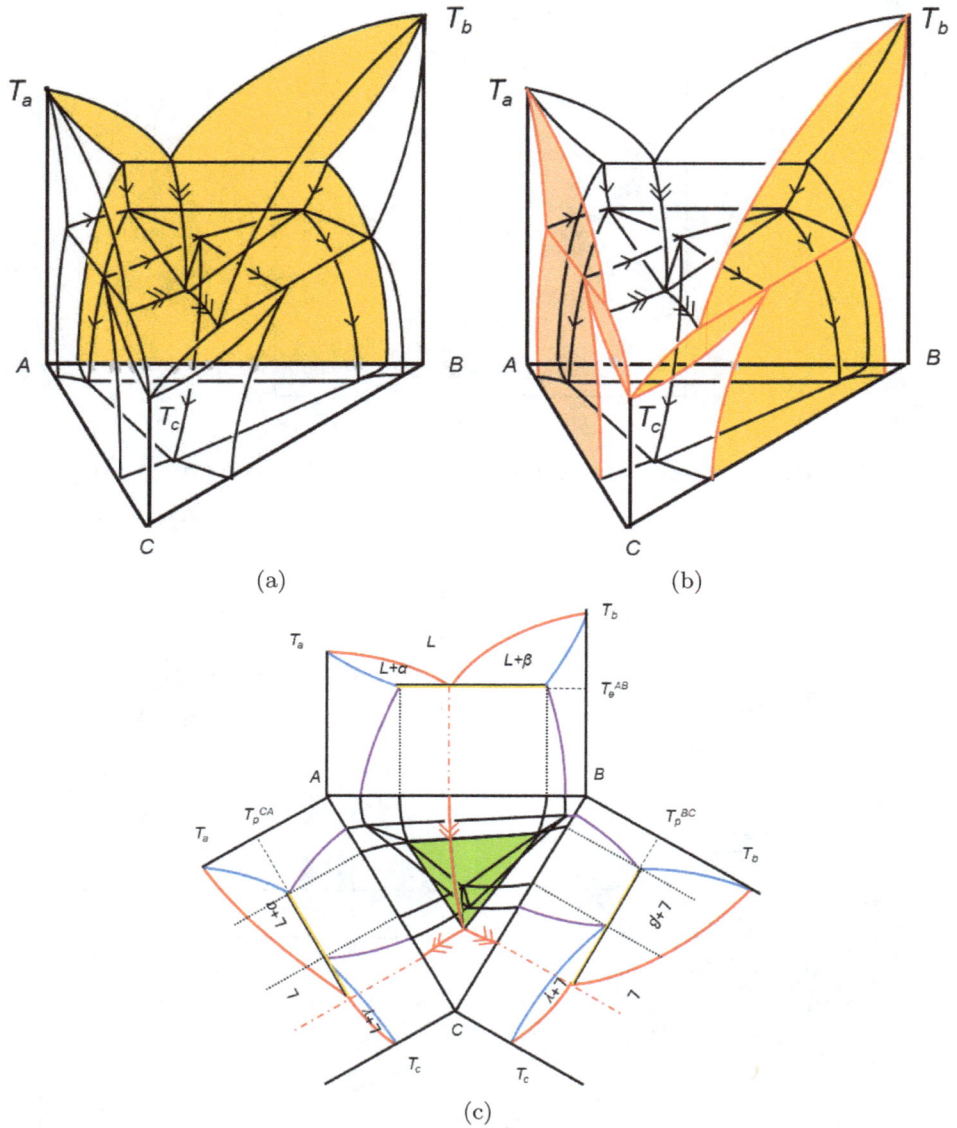

Fig. 4.49 Peritecto-eutectic system (Class III). (a) Shaded area indicates A-B binary eutectic system. (b) Shaded areas indicate B-C and A-C binary peritectic systems. (c) An opened-up diagram.

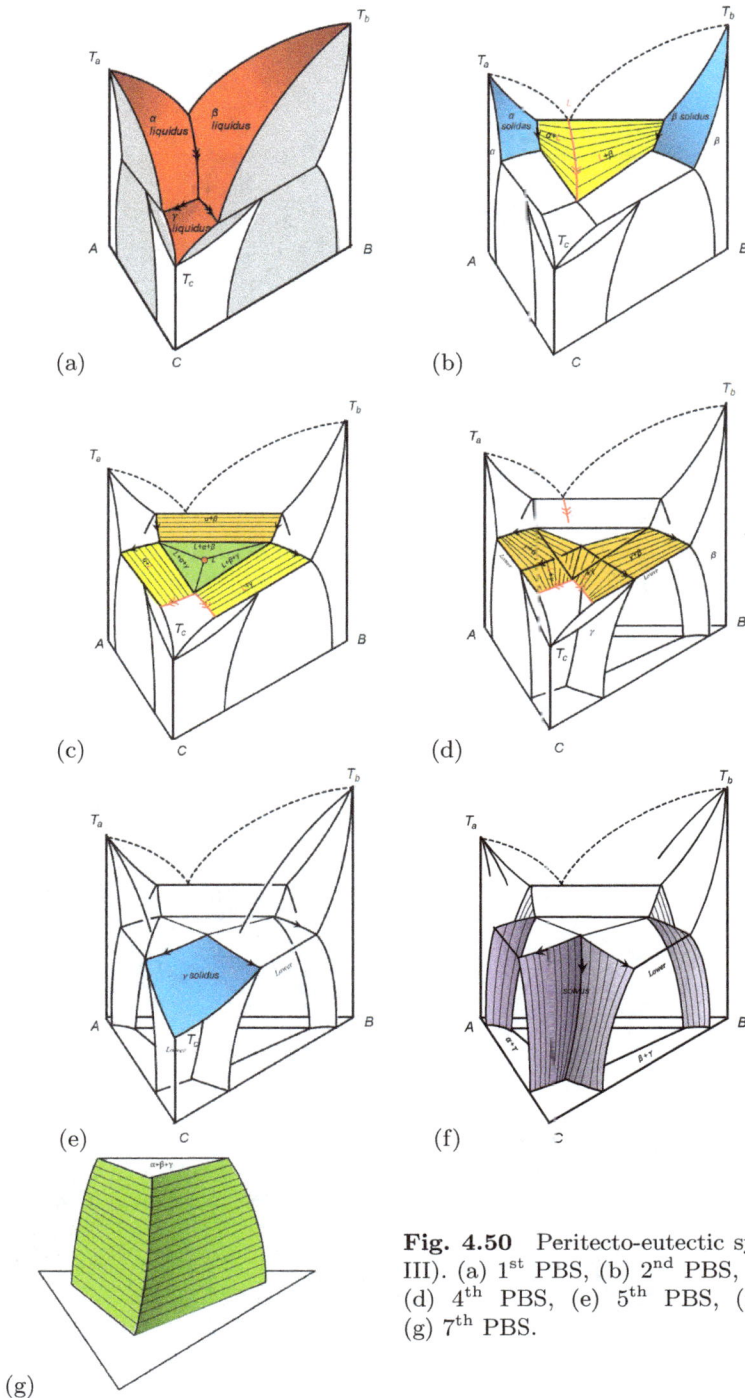

Fig. 4.50 Peritecto-eutectic system (Class III). (a) 1st PBS, (b) 2nd PBS, (c) 3rd PBS, (d) 4th PBS, (e) 5th PBS, (f) 6th PBS, (g) 7th PBS.

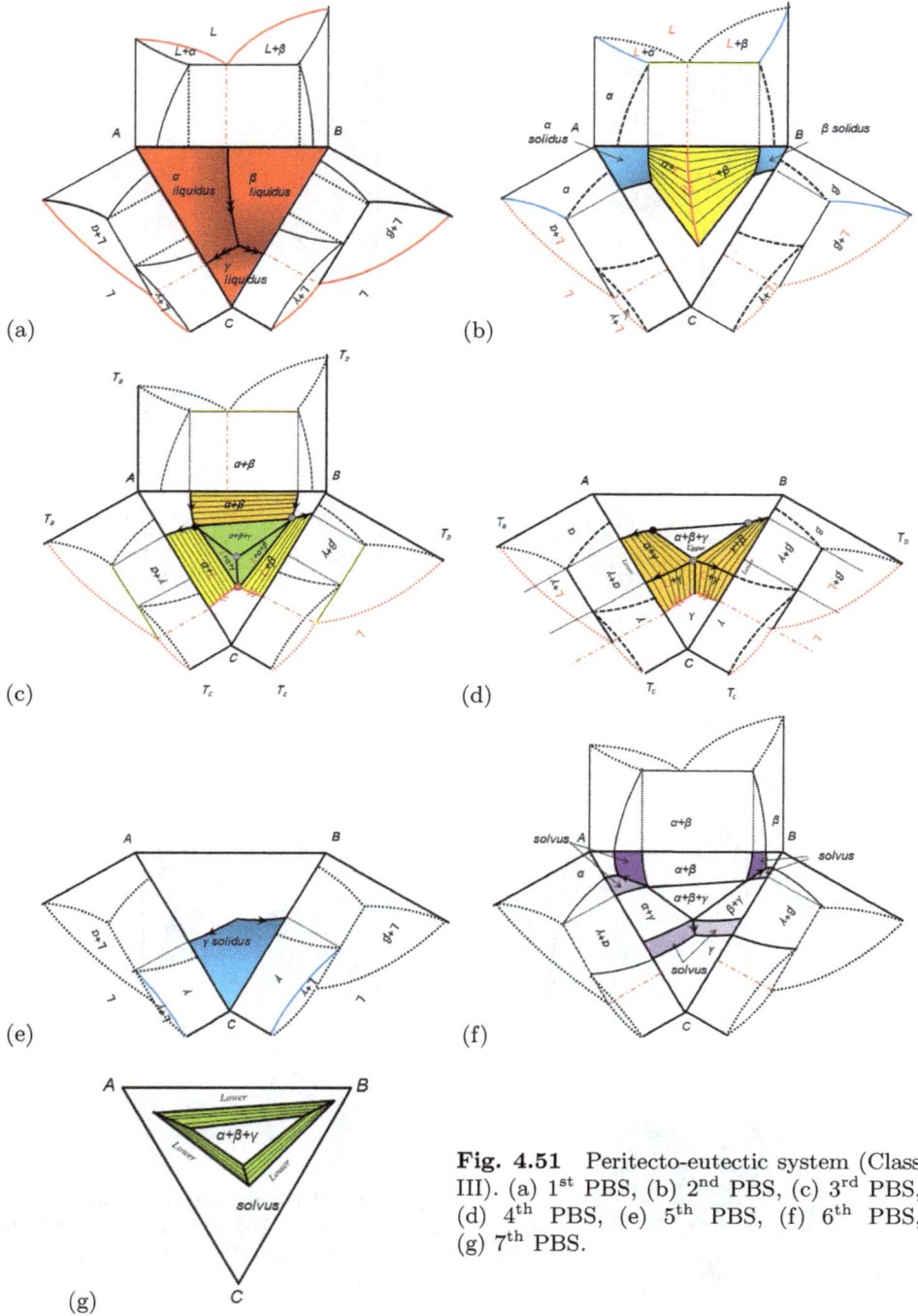

Fig. 4.51 Peritecto-eutectic system (Class III). (a) 1st PBS, (b) 2nd PBS, (c) 3rd PBS, (d) 4th PBS, (e) 5th PBS, (f) 6th PBS, (g) 7th PBS.

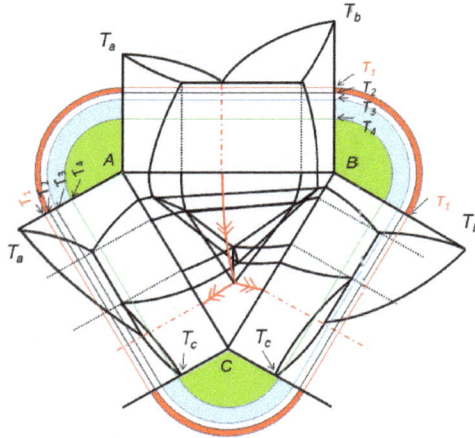

Fig. 4.52 Drawing isotherms at T_1–T_4 (Class III).

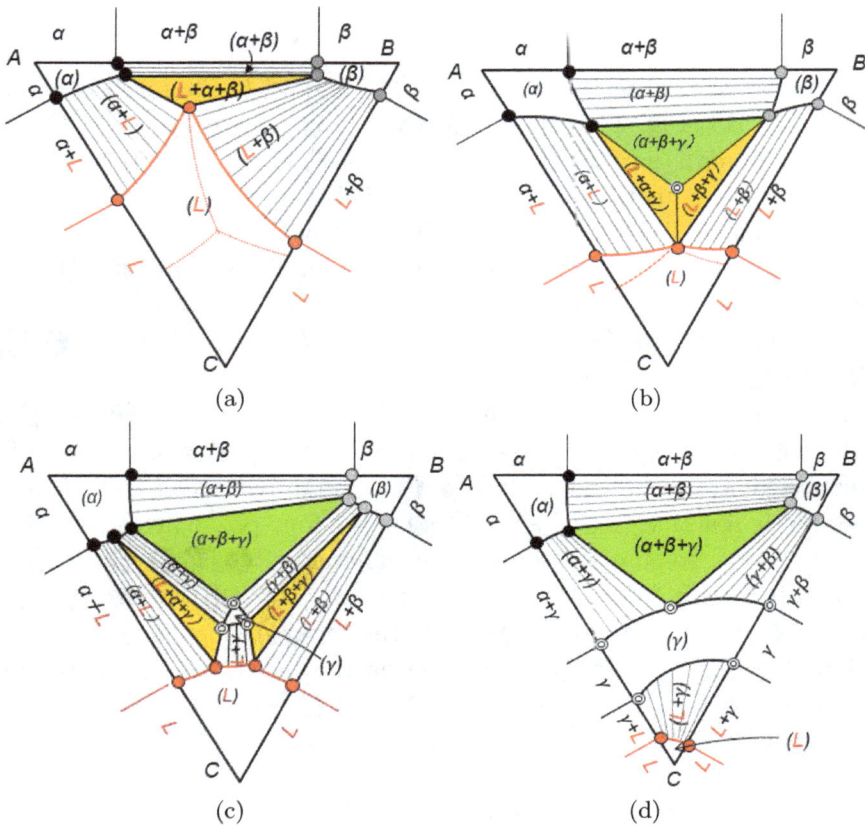

Fig. 4.53 Peritecto-eutectic system (Class III). Isotherms at (a) T_1, (b) T_2, (c) T_3, (d) T_4.

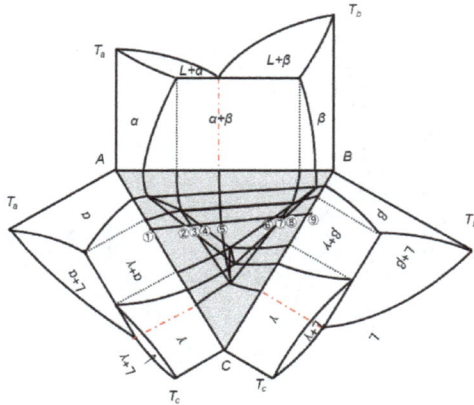

Fig. 4.54 Drawing a vertical section along ①–⑨ (Class III).

b. Vertical sections

Let us draw a vertical section along ①–⑨ in Fig. 4.54. Characteristic points on the first PBS (Fig. 4.55(a)) are ❶, ❺ and ❾, where ❶ = ①′, ❾ = ⑨′. ❶–❺ is the liquidus surface of α-phase and ❺–❾ is the liquidus surface of β-phase. Since the eutectic reaction in the A-B binary system is accompanied, ❶ ◱ ❺ ◲ ❾ with the liquidus line ❺ at valley (bottom).

Characteristic points on the second PBS (Fig. 4.55(b)) are ❸, ❺ and ❼. ❸–❺ is the roof $(L + \alpha)$ of the eutectic reaction $(L + \alpha + \beta)$ in the A-B binary system, and ❺–❼ is the other roof of $(L + \beta)$. Therefore ❸ ◲ ❺ ◱ ❼ with ❺ at the summit.

Characteristic points on the third PBS (Fig. 4.55(c)) are ❶, ❸, ❼ and ❾, where ❶ = ①″, ❾ = ⑨″. ❶–❸ is the upper broad face $(L + \alpha)$ of the peritectic reaction $(L + \alpha + \gamma)$ and ❼–❾ is the upper broad face $(L + \beta)$ of the peritectic reaction $(L+\beta+\gamma)$. ❸–❼ is the four-phase invariant reaction $(L + \alpha + \beta + \gamma)$, so that it is horizontal. Therefore, ❶ ◲ ❸–❼ ◱ ❾.

Characteristic points on the fourth PBS (Fig. 4.55(d)) are ❶, ❹, ❻ and ❾, where ❶ = ①″, ❾ = ⑨″. ❶–❹ is the lower narrow face of peritectic reaction $(L+\alpha+\gamma)$ and ❻–❾ is the lower narrow face $(\beta+\gamma)$ of the peritectic reaction $(L + \beta + \gamma)$. Therefore, ❶ ◲ ❹ and ❻ ◱ ❾.

The fifth PBS (Fig. 4.55(e)) has no characteristic points, and is ignored. So, is the sixth layer (Fig. 4.55(f)).

Characteristic points on the seventh PBS (Fig. 4.55(g)) are ❷, ❹, ❻ and ❽, where ❷–❹ and ❻–❽ are the walls of the pyramid of the three-phase triangle $(\alpha + \beta + \gamma)$, so that ❷ ◲ ❹ and ❻ ◱ ❽.

The resultant vertical section along ①–⑨ is shown in Fig. 4.56.

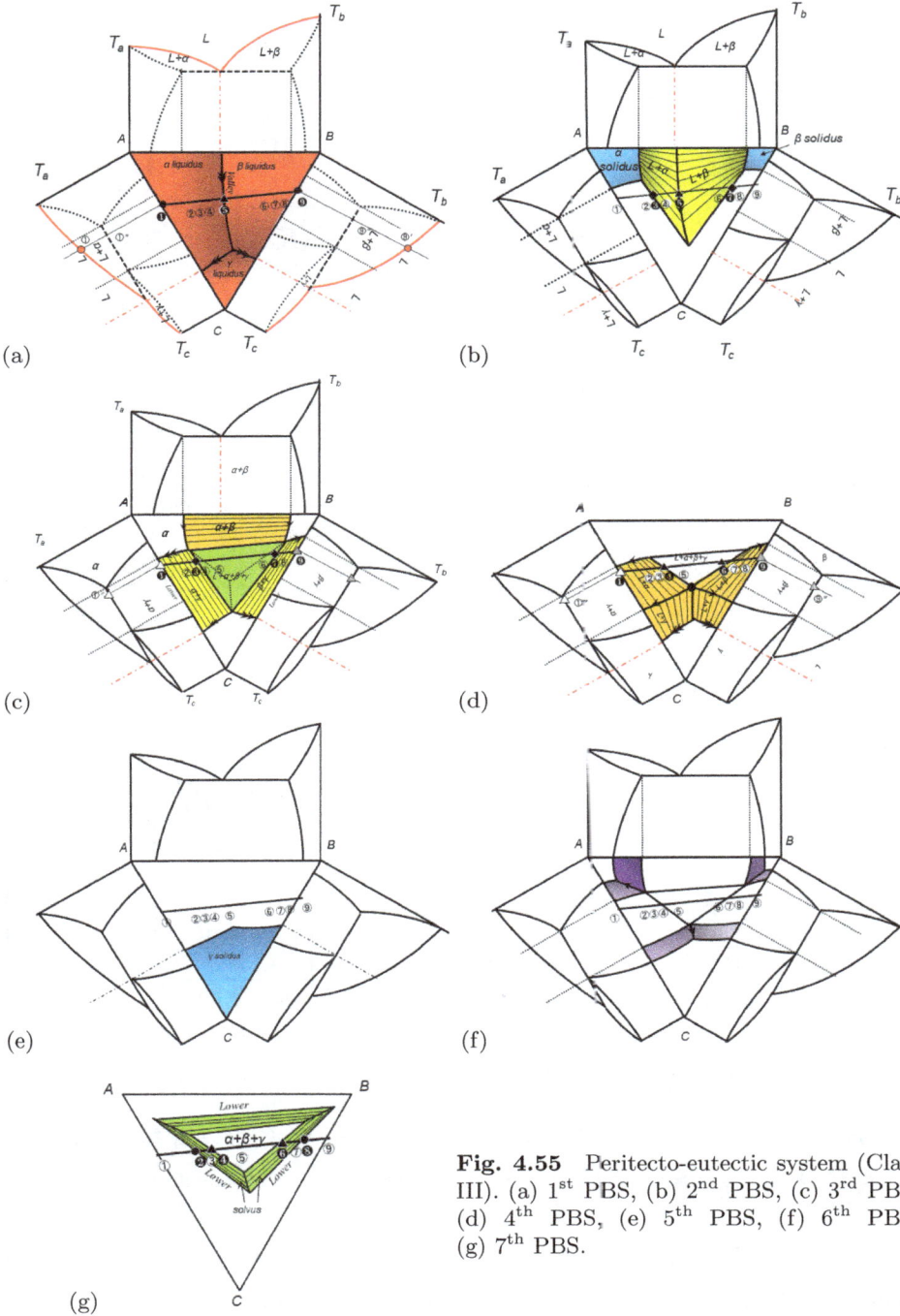

Fig. 4.55 Peritecto-eutectic system (Class III). (a) 1st PBS, (b) 2nd PBS, (c) 3rd PBS, (d) 4th PBS, (e) 5th PBS, (f) 6th PBS, (g) 7th PBS.

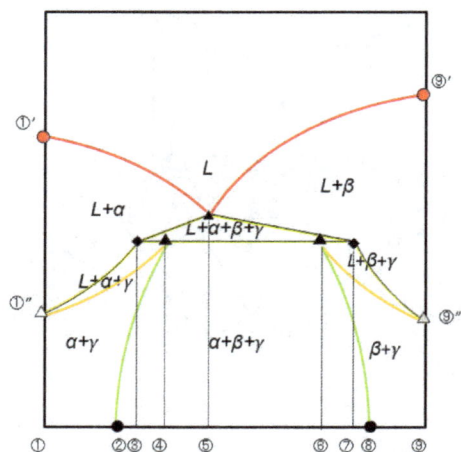

Fig. 4.56 A vertical section along ①–⑨ in Fig. 4.54 (Class III).

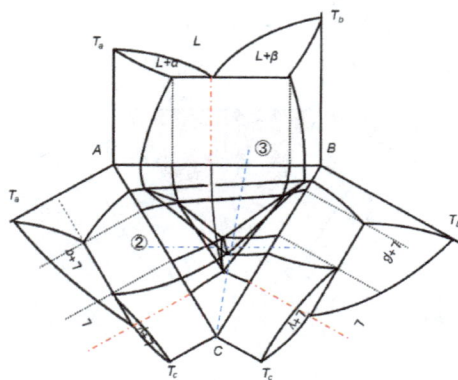

Fig. 4.57 Drawing vertical sections along ② and ③ in Fig. 4.53 (Class III).

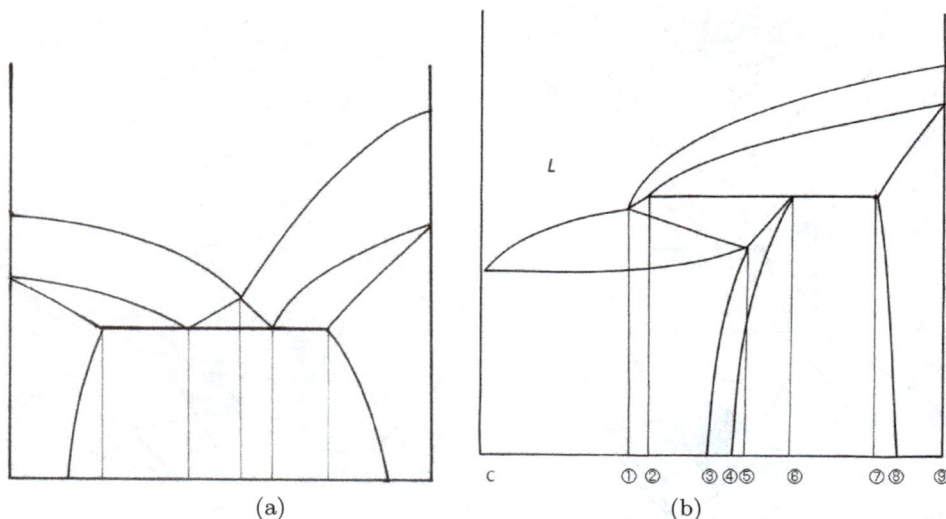

(a)   (b)

Fig. 4.58 Vertical sections along ② and ③ (Class III).

Exercise. Draw vertical diagrams along ② and ③ in Fig. 4.57. (Solutions are shown in Figs. 4.58(a)(b).) Name the phases in each of the fields.

4.3.4   Summary

When all of the A-B, B-C and C-A binary systems have invariant reactions, in the A-B-C ternary system, three-phase triangles $(L+\alpha+\beta)$, $(L+\beta+\gamma)$ and $(L+\gamma+\alpha)$ join to form $(L+\alpha+\beta+\gamma)$ invariant reaction. If L is inside the

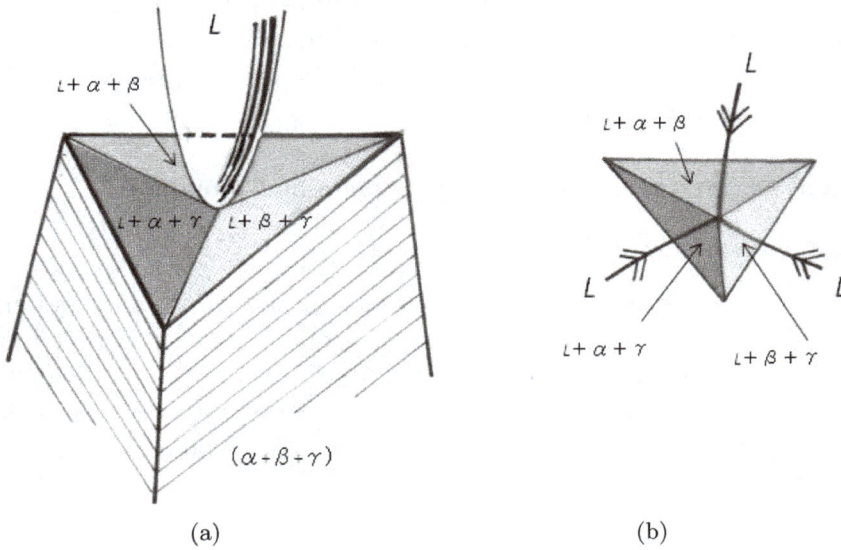

Fig. 4.59 Relationships between L and $(\alpha + \beta + \gamma)$ three-phase triangle in a ternary eutectic system. (a) Perspective and (b) Plan view.

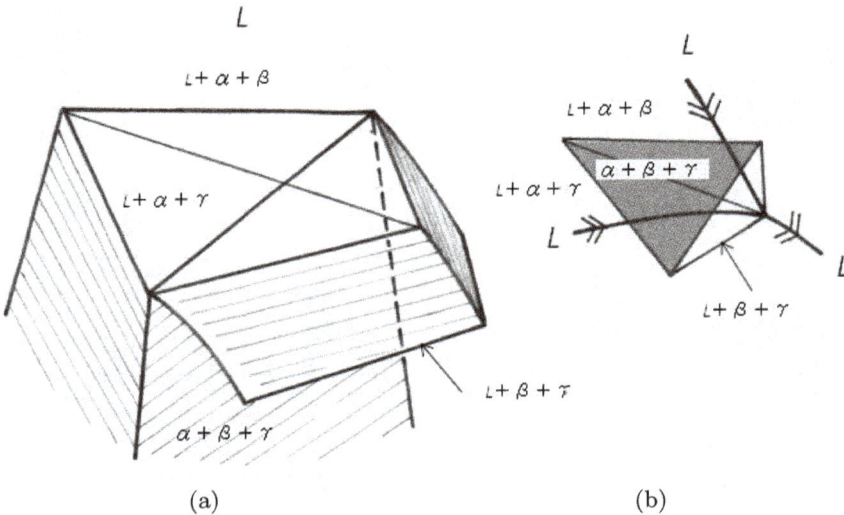

Fig. 4.60 Relationships between L and $(\alpha + \beta + \gamma)$ three-phase triangle in a ternary peritecto-eutectic system. (a) Perspective and (b) Plan view.

three-phase triangle $(\alpha + \beta + \gamma)$ (solid), this results in the ternary eutectic system (Fig. 4.59). Since L is the liquid phase, it disappears on lowering the temperature, with the remaining solid three-phase triangle (pyramid) descending perpetually to room temperature.

By contrast if L is outside the three-phase triangle $(\alpha+\beta+\gamma)$ (solid), this results in peritecto-eutectic reaction (Fig. 4.60). Both L-containing three-phase triangle and solid three-phase triangle $(\alpha + \beta + \gamma)$ descend, but the liquid L descends much more slowly.

4.4   Ternary Systems Containing Intermetallic Compound

4.4.1   Quasi-binary system

Suppose that a compound A_mB_n is present in the A-B binary system as shown in Fig. 4.61. When the quasi-binary reaction is present between A_mB_n and C as shown in Fig. 4.62, the A-B-C ternary system can be divided into two independent A-A_mB_n-C and A_mB_n-B-C quasi-ternary systems as shown in Fig. 4.62(a). Figure 4.63(a) shows the valley of the liquidus surface with an arrow.

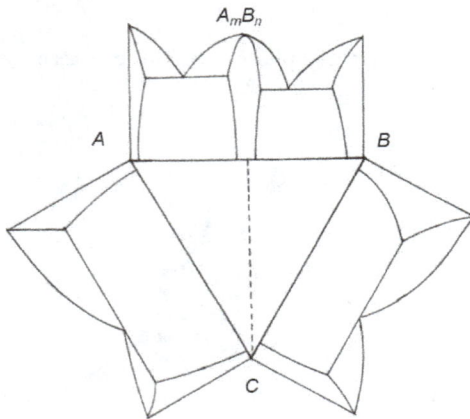

Fig. 4.61   When a congruent intermetallic compound A_mB_n is contained in the AB binary system.

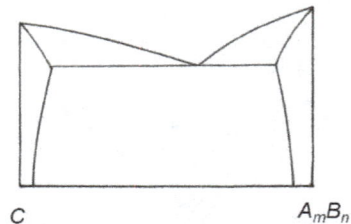

Fig. 4.62   Pseudo-binary system between A_mB_n and C.

4.4.2   Quasi-binary systems are not formed

When the liquidus surfaces of either α of the A-A_mB_n-C system or β of the A_mB_n-B-C system flow into the neighboring A-A_mB_n-C and A_mB_n-B-C systems as shown in Fig. 4.63(b), neither A-A_mB_n-C nor A_mB_n-B-C can be treated as independent quasi-ternary systems.

A more complex system is shown in Fig. 4.64, where the valley of the liquidus surface passes through the peritecto-eutectic reaction, then P_1 passes P_2 and eventually reaches the ternary eutectic reaction at E.

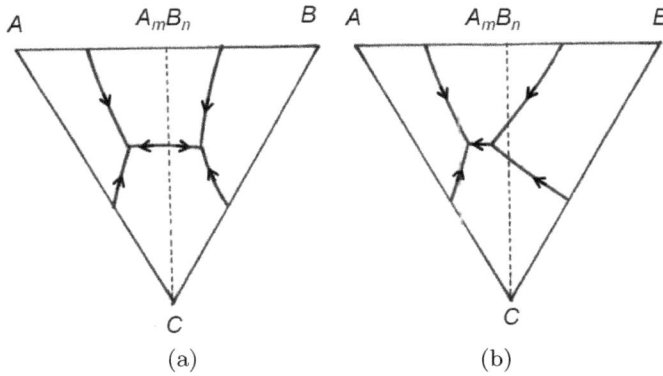

Fig. 4.63 A-B binary system contains a congruent intermetallic compound A_mB_n. (a) When a pseudo-binary system exists between an intermetallic compound A_mB_n and C. (b) When a pseudo-binary system does not exist between an intermetallic compound A_mB_n and C.

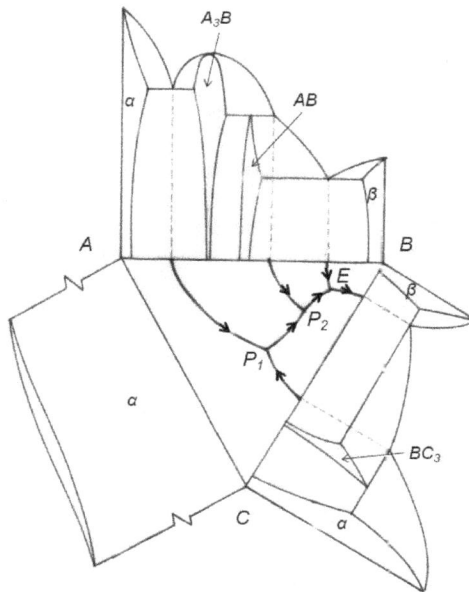

Fig. 4.64 A complex ternary system. Arrows indicate flows of valleys of the liquidus (L).

Appendix: Conversion of indices between three and four indices in HCP

Indices based on the three axes, a_1 and a_2 and c are generally used for hexagonal crystals but are open to the objection that equivalent planes do not have similar indices. For instance, the planes (100) and $(\bar{1}10)$ are both prismatic planes and are equivalent. The same objection applies to the indices of directions. For these reasons the hexagonal indices such as those shown in Fig. A.1 (same as the figure in footnote on p. 75) are used. In addition to the usual a_1 and a_2 axes, a third axis a_3 is introduced and planes and directions are indicated by $(hkil)$ and $[uvtw]$, respectively. Here, h, k, i or u, v, t are indices corresponding to the a_1, a_2, and a_3 axes. l and w are indices corresponding to the c axis. h, k, i and u, v, t are related by the equations

$$h + k = -i, \qquad u + v = -t \tag{A.1}$$

Transformations between the three and four-indices, i.e., between (HKL) and $(hkil)$ and between $[UVW]$ and $[uvtw]$ are given by

$$H = h, \qquad K = k, \qquad L = 1, \qquad i = -(h+k) = -(H+K) \tag{A.2}$$

$$U = 2u + v, \qquad V = u + 2v, \qquad W = w$$

$$u = 1/3(2U - V), \qquad v = 1/3(2V - U), \qquad w = W, \qquad t = -(u+v). \tag{A.3}$$

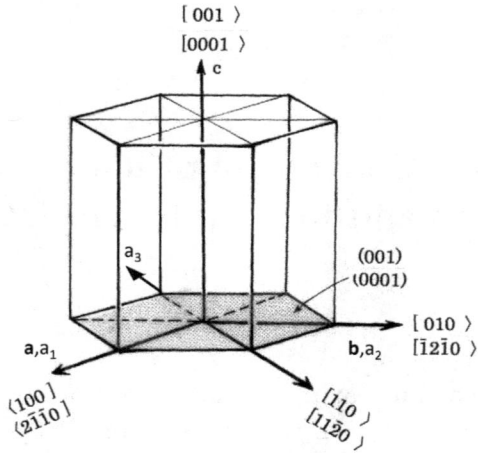

Fig. A.1 Coordinates of **a**, **b** and c are for three-index system. Coordinates of a_1, a_2, a_3 and c are for four-index system. ( ) stands for plane and [$>$ or $<$] stands for orientation.

Index